1st Edition printed in 2000.

2nd Edition completely revised and amended into 4 books:

Book 1 - ISBN 9781448610068 - 2010
Book 2 - ISBN 9781466405530 - 2011
Book 3 - ISBN 9781466416161 - 2011
Book 4 - ISBN 9781466424609 - 2011

Published by Susan Munzer

This book is the third out of a series of four books by the same author.

For more information visit our website:
www.learn2play2learn.com

Dedication

To my parents, Werner and Hanna Briner, who have given their three daughters the best of their heart and strength. My Mom, who has encouraged me to be creative and my Dad, who taught me to be practical and clever with my hands, as well as my two sisters who were part of a great childhood.

To my family, Eric, Marc & Nisha for sharing all these wonderful years. For Eric whose support and dedication to the family has made our lives special and for Marc and Nisha for giving me the opportunity to live a dream and see it fulfilled.

To Oliver, Daniel and Josefine, my first grandchildren and Christina and Dean who have joined our family.

To early childhood education professionals around the globe, who work with dedication and selflessly, and often without adequate compensation, to give children a good foundation in life.

Special Thanks To:

Nisha, for typing the entire book and editing my "Swiss" English text and for being there when I needed her help and encouragement.

Marc, for helping with the creation of the second edition of this book series and for being my special advisor, whose computer expertise, editing help and belief in this book series has been invaluable.

Stuart Dolling, the electronic publishing professional for his valuable help.

All the people who throughout the years have helped me to refine and develop my work through interesting discussions and constructive feedback.

All the wonderful and interesting people I have met through my workshops, whose feed-back has encouraged me to pursue this book series.

The Learn to Play - Play to Learn Series

This book and DVD series is a set of four illustrated practical guides for parents and caregivers of small children, which examine the relationship between playing and learning and how we as parents and caregivers can support the children in their learning processes.

Have you ever wondered how much children learn as they play, or how important it is for their development to have good toys and the right toys? Did you know how damaging bad toys are, and how easy it is for you to learn to make the right choices?

If children play with the right toys in their very early years, it will get them started on the right path in life. As they learn mental quickness, self-esteem and self-control, to mention just a few, they are later able to master their school years with more interest, and they will have the right foundation for learning.

These are just some of the topics which I am exploring in this book and DVD series. At the same time, I am sharing more than 100 of my self-made toys and 42 story baskets of which 32 I have especially written and created for this series. There are over 450 photographs and 70 patterns with how-to-do explanations. They have made a difference in my children's lives and in the lives of many children of caregivers and parents who I was privileged to teach over the last 40 years. They can make a difference in your child's life, too.

I hope that this series will help you become more creative and more flexible in your teaching style as a parent or as a teacher.

I hope that this series will become a tool for you to teach foundational concept learning that will let children put real building blocks of learning into their development.

I hope that this series will allow you to feel the excitement of knowing that you were instrumental in giving a child a good start in life.

The book and DVD series comprises four books and four DVDs all separately available. The books and DVDs are described in more detail on the next page.

Further educational materials which support the methods and concepts in the Learn to Play - Play to Learn series are available through our webpage:

www.learn2play2learn.com

This series is made up of four books. Each part comprises both a book and a DVD film. The book explains the concepts with pictures, patterns and text and the DVD film supplements the book by illustrating the concepts via a filmed presentation with the author.

The four books are:

- Book 1 - Educational Storytelling
- Book 2 - Story Baskets
- Book 3 - Math and Science
- Book 4 - Puppets and Empathy

Book 1 - Educational Story Telling

This book looks at what children learn as they play and as they hear stories. We look at questions like: How do we bring out childrens individuality, potential and talent? How do we reduce TV time and fill it with great learning? This is a great parent and teacher support and provides ideas which are much loved by the children!

Book 2 - Story Baskets

In this book we introduce 25 story baskets in words and pictures. With the Story Basket method we learn to be flexible, use any materials at hand to entertain, delight and teach children. Much effort is put on virtues and empathy which can be used to provide great lessons on good morale. This is an all time childrens favorite!

Book 3 - Math and Science

In this book we introduce ways of teaching math and science by concept learning in order to build a solid foundation for later school years. We discuss how to choose and make toys and exercises for hands on learning with easy to find materials. Discover nature as a great teacher. Included are 9 story baskets focussed on math and science.

Book 4 - Puppets and Empathy

In this book we look at puppets and how they can be used in education. Children can easily connect and relate to puppets which allows puppets to present amazing opportunities to teach virtues and empathy. We look at different varieties of puppets and show how to make some yourself. Included are 7 more story baskets focussed on empathy.

All four books include do-it-yourself projects with patterns and helpful tips.

Math and Science
TABLE OF CONTENTS

Math and Science

Story Baskets:

2 × 2 + 6 Goldilocks
- 3 Goldilocks
= 7 Goldilocks

5 + 8 Bears - 3 Bears
× 3 Bears + 8 Bears
-2 Bears
= 36 Bears

INTRODUCTION

Math and Science — How Exciting!

In a small child's world it is hard to separate math and science. We are surrounded in everyday life by concepts of both these subjects. There are so many things for us to point out, to look at, observe and study, which make life very interesting, and lay that foundation in a very easy way.

When we introduce math and science into a child's life early in a playful manner, it becomes second nature, and we teach that it has an important function in our life. It is not just a bother that is taught in school.

It is true that math and science can be hard for some children, because they have very different interests and fields of understanding. However, it is very important to plant the knowledge for conceptual understanding early on, so they can develop the skills for math and the confidence that it is not difficult to succeed.

There are patterns of math all around us, in music, rhythm, plants, animals, rock formations, semi-precious stones. You can recognize symmetry, spirals, rosettes, crystal forms and mineral structures. You find all shapes in nature, circles, triangles, hearts, squares, ovals and cubes.

It is all very fascinating. For example, crystals always grow according to simple mathematical laws. You can study the crystal geometry and find that every crystal will fit into one of the following six categories: isometric, tetragonal, hexagonal, orthorhombic, monoclinic and triclinic.

I usually like to try to keep my book on a level of understanding that is easy to grasp. A fact like the one above shows us that there is a marvelous system in everything, if we dig a little deeper. You can look at a nature book with your children and let them find for example, symmetry, spirals or circles etc. And then of course point them out in nature, in the bath, on their own bodies, or looking in the mirror.

Have you ever seen pictures of crop circles, the mysterious designs in crop fields that baffle our intellect? This all makes us aware of a glorious design in nature that we come across everywhere in our daily life.

If we make a point of seeing the hidden, the small detail, the patterns of the seasons, we open up an amazing world for the child's everyday life. To learn to observe, contemplate and develop an understanding can be part of a solid foundation in the growth of a child's development.

They can be amazed, surprised, happy, touched, and intrigued and so build an interest, an excitement, learn to be inquisitive and have their eyes open as they walk through life. This is real learning.

In my teaching years, I have designed and made many toys and written story baskets to supplement the themes and different stations for the children at the centers. I was also able to use my curriculum materials as hands-on examples for ECE personnel and students, in my presentations and workshops.

It has been very satisfying to see the effectiveness that educational, hands-on materials can give in these early years for children and in support for the teaching staff.

I have chosen a selection of these projects for Book 3 Math and Science that are typically related to these subjects. However, many of my other ideas and projects in Books 1, 2 and 4, also relate to Math concepts or a process of nature. This is the wonderful thing about teaching children—the world is their classroom, and many subjects are mixed in with each other to connect the different parts and build a whole for the child.

Whether you believe you
can do a thing or not,
You're right.

Henry Ford

Math Concepts

There are many different learning styles for math and I would like to discuss memorization versus concept learning. From early on we have a tendency to teach math by way of memory. We flash numbers and answers at the children for them to memorize, instead of making sure that they have grasped the concepts. Sooner or later they will have a difficult time understanding when memorization becomes too difficult.

If you thought that children were just playing, then I hope I'll be able to change your mind with this chapter.

During our son Marc's university studies in Engineering, it was always his goal to understand every concept. He was only satisfied if he could get to the root of the problem and understand all the different steps. Most of his classmates would work through memorization. They would memorize certain steps from one example and apply it to another problem. Marc always wanted to understand each step and then work through problems that way. If the teacher would allow cheat-sheets, Marc would just put a few guidelines for each step while some other students would cram as much as they could onto one page.

I wonder how a style of learning affects a school, country or industry?

For very small children, I believe strongly that hands on concept learning is essential and easy to do by exposing the children to the right toys. The earlier you train them to have the right thinking processes, the better. In my earlier teaching years, I was always very frustrated with the parents of preschoolers (3 and 4-year-olds). They always thought that if the child didn't learn numbers and letters, the preschool was no good. Finally, I designed a "toy for the parents" where I could show what children learn as they play. It was such a hit and right away the parents realized what good toys can do. I would show them the different learning centers and explain what each one was able to teach. They got quite excited! We would always have corners set up with early science, literature, dramatizing, art, math, music etc. For me, setting up these different areas was always a joy. It was very satisfying to watch the children playing or working in the different areas. I would almost get overwhelmed with excitement when I realized how their minds were working and all the new things that they were able to learn.

If we expose children to a lot of hands on concept learning, we do not have to teach them numbers and letters very early. We can wait until they show interest themselves. When they come to Kindergarten (five-year-olds), the foundation for learning will be set and learning will be easier.

You will find the Math Puzzle Box "for the parents" on Page 17.

HOW TO MAKE PUZZLES

You can see some of my puzzles in Book 1, 2 & 4 also, but in this chapter I would like to show you how to make them. Some puzzles have a science or math theme where children can learn about different stages of growth, new life and changes in a season as well as counting, colors and shapes. Other puzzles have more of a story context, and are connected to a song, a poem, the family or a special celebration.

Puzzles with layers are very versatile. You can give a child only one layer to work on and exchange layers as the child masters the previous one or you can add a layer to the one that the child has already played with. When you are working with small children and have a class of 3-year-olds and a class of 4-year-olds, you can take away a layer for the three-year-olds. You can actually use the same puzzle for children at many different ages. The frog, the duck, the butterfly and the apple puzzles all have the same sized squares as the three layer egg puzzle or the three layer animal puzzle, having twelve squares each. You can take three squares from either the animal puzzle or the egg puzzle and put them into one of the smaller boxes. This would be an ideal beginners puzzle. Or, you can put the butterfly, the frog, the duck and the apple all together in a larger box and have a very difficult puzzle for a five or six-year-old.

In the twelve square animal box or the twelve square egg box, every square has only three pieces, so every individual puzzle is very easy. The whole box has 36 pieces, so putting it together is still an accomplishment. By matching the color schemes, the children can find which pieces go together. They learn that a big job can be done in small steps. These puzzles are built on the concept of work and reward. As the children finish the puzzle, they get spurred on by the feeling of accomplishment. One large puzzle with just one layer seems to be an incredibly big job and more discouraging for smaller children.

I have made two different frog puzzles. You can see that with all puzzles you can cut one design in many different ways. The easier one has 15 pieces and the harder has 30. The easier one would be suitable for a three-year-old, and the harder one for a four-year-old. The three-year-old would then not always be the one to finish last. Many times the younger children are finished last and work a lot slower. This can become discouraging to the point where they believe that they are not as good. The role of a second born is different and we should not worry too much about situations such as this, but it would be nice to make the children aware that developmental steps are different in every age. Two puzzles done in this way could bring such awareness.

These puzzles are not difficult to make, but they do take many hours. To make them and sell them is almost impossible. You would have to charge too much and then nobody could afford them. You would have to go into production in a big way to make it worthwhile. However, for your own children, for your childcare center, or as gifts, it is always worth it.

A life spent making mistakes is not only more honorable, but more useful than a life spent doing nothing.

George Bernard Shaw

Puzzles made by 7 to 10 year old children

You Can Do It Too!

As you look at my puzzle section, you might think that you could never make these yourself. I hope to show you otherwise. I have often made puzzles with children and they have always been able to complete them. You don't want to start with the more difficult ones if you do not have much experience. Choose an easy one, preferably something that you can draw yourself and that is just one layer. In my early years, I used a fret saw which is a hand saw designed for plywood. If you look in your grandpa's workshop he may still have one. The saw blades are very thin, almost like wire with teeth, so that you can make very sharp turns. You can practice first on a fret saw. You might find it difficult at first but you will catch on. The secret is to keep steady pressure moving forward, otherwise, you will get stuck and break the saw. To work on a small, electric table scroll saw is easier than working on a sewing machine. Some saws have a built-in vacuum so there isn't much sawdust and you can actually use them in the house. Remember practice makes perfect. There is more information on saws in the chapter on tools and materials.

Puzzles made by 5 year old children

How to Make a Simple Puzzle

Take one piece of 1/4” (6 mm) plywood and another piece the same size, but 1/8” (3 mm). Cut a piece of paper the same size to use as a pattern. You can either draw a frame first on your sheet of paper and use the whole square inside as your puzzle, like the boat puzzle. For example, if you design a butterfly, the whole background of the butterfly can be a frame. When you design a puzzle you have to remember that you are working with wood. Keep in mind that the pieces cannot be too small, too narrow or too pointy. You don’t want any pieces to break off or get lost easily.

By designing your own puzzle, you can learn the most. Right away you will see what works and what doesn’t. Once you have the pattern and know how you are going to cut it, you can transfer the pattern by carbon paper onto your 1/4” (6 mm) plywood. Then you make a drill start. Since you do not want to just cut into the frame, you want the frame to stay in one piece. Like this, you can cut out the inside piece. (See next Page).

When you cut out the frame, cut out one of the corners visibly different from all the other corners so that the children know how to start the puzzle. Because it is hand made you will never be able to get it exactly straight which results in the puzzle fitting in only one way. Once you have the frame, you sand the inside really well and glue it onto your 1/8” (3 mm) piece of plywood. Sand the outside when the two pieces are together.

Now that you have the bottom part finished, you are safe to cut the puzzle into pieces. As you cut up the pieces, you can put them into the frame, and use it as a tray. I always keep the puzzle together as I work on the different steps. It is difficult to put the puzzle together when the pieces are not painted.

Sand all the individual pieces very well, especially the edges and their corners. If a piece of wood breaks off, glue it back on, let it dry and sand back over it. I use fine sandpaper (number 150). The better a job you do on your sanding, the better the finished toy will look. Your puzzle is ready to paint now. There will be more on paints in the materials and tools section.

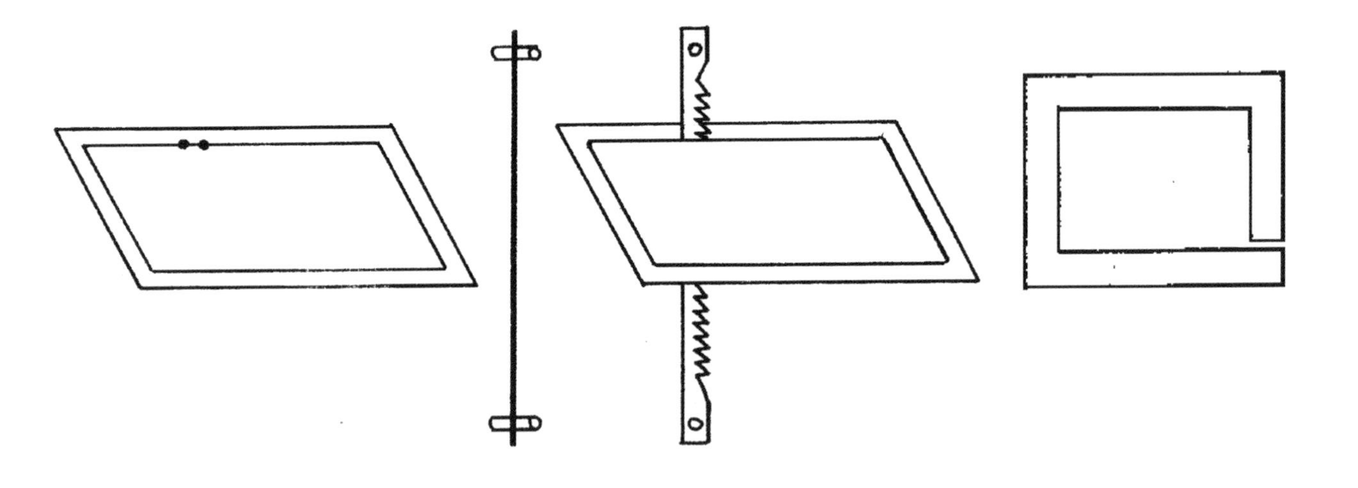

Drill Start - What it is and How to do it

When you make a puzzle where you need a frame, you don't want to cut through the frame, because it weakens the puzzle and doesn't look nice. So you make a drill start. If you have a lot of space, as with the boat in the story of "Peter the Fisherman" (See Photo in Book 2, page 29), you can drill a fairly large hole to fit your saw blade through. With my saw blade a 5/32" (4 mm) drill will be enough. If you need the inside of the frame for your puzzle you want to drill two very small holes that you connect so you will be able to insert the blade. With my saw it is a little complicated, the saw blade has little pins on both sides to attach to the saw. I have to remove one of the little pins, stick the blade through the little slot in the wood that I made with the two little holes. Then put back the little pin and fasten the blade with the wood piece attached to it to the saw. Now you can cut out your middle piece. It sounds difficult but it works. Make sure you have the pattern on top of the wood and the blade teeth pointing down.

How to Make Boxes

In the beginning I made very simple boxes.

<u>You need</u>: 1 bottom part - piece of 1/8" (3 mm) plywood;
4 pieces of molding for the sides

Fig. 1 & 2 show you a simple way of putting together a box. You can cut the pieces of molding with a hand saw. If you want to get more sophisticated you can use a mitre box and make corners as in Fig. 3. I can assure you that children will not see a difference and they will be just as happy and learn just as much with a simpler box! When all the pieces of moldings are cut, glue them on the bottom sheet and hold them in place with an elastic. Make sure you put glue into the corners between the moldings. When it holds together somewhat, turn it around and put two finishing nails on the bottom for each molding. Cut all the squares that go into the box, sand the box and all the squares. Now the box is ready and you can trace the patterns on the squares, then finally cut the squares according to the pattern.

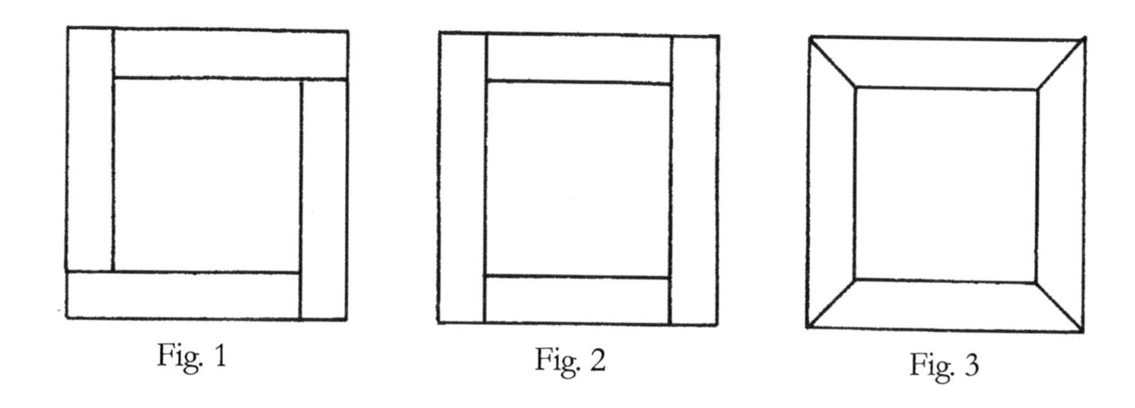

Fig. 1 Fig. 2 Fig. 3

Moldings for Boxes

I use different types of moldings for the boxes. They are a good quality wood and there is a good variety. You will have to see what you can find at your lumber store. I mainly use three different types. Choose good strips without cracks or slivers.

Make sure you design your boxes a little larger than your squares. I like to have about 1/8" (3 mm) extra space. If the pieces fit too tightly into the box, it can be frustrating for the children. If there is too much space the puzzle doesn't sit nicely.

<u>Box 1</u>: This molding fits nine layers of plywood. As you can see it is rounded on one corner, and makes nice boxes.

<u>Box 2</u>: This one fits five layers and is also rounded. I cut this one down for my "Four Seasons Apple Tree". The height of this puzzle is three layers of 1/4" (6 mm) ply wood plus one layer of 1/8" (3 mm). For the "Christmas Tree Puzzle" which is two layers of 1/4" (6 mm) plywood plus one layer of 1/8" (3 mm), I cut it down accordingly.

<u>Box 3</u>: This molding is ideal for three layer puzzles and for larger boxes. I use this one for the hammering game also.

This drawing shows the profile of the moldings and the layers in a box.

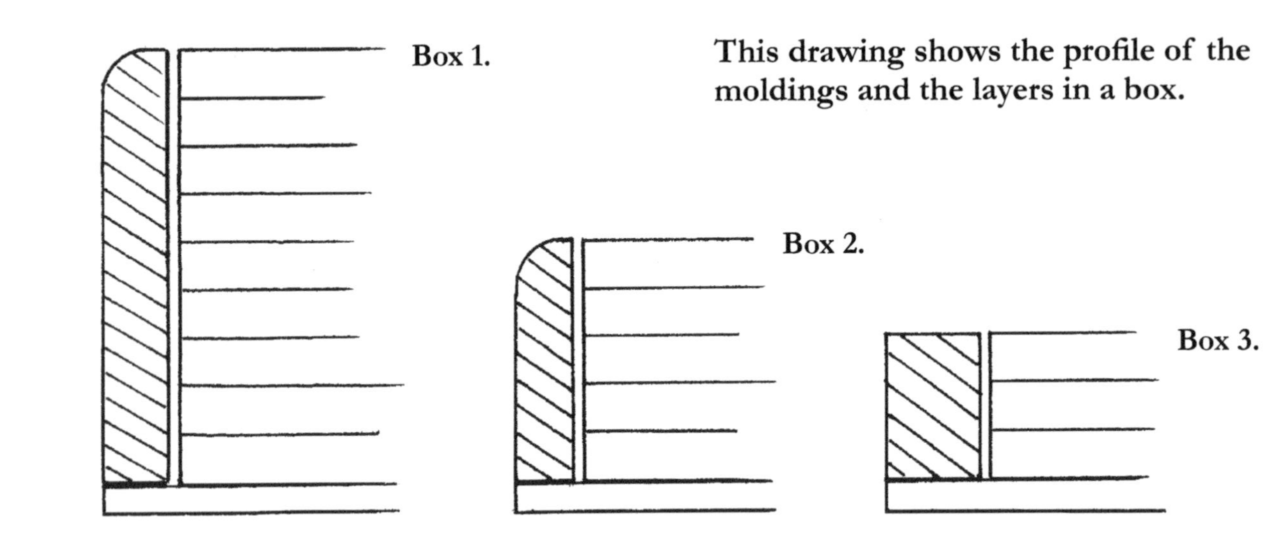

MATH PUZZLES

The Nine-Layer Math Puzzle Box

The nine-layers are all identical squares cut from 1/4" plywood. Each square is cut up into geometrical shapes and painted with one color, so the children know which pieces belong to which square. I usually use bright colors for each square. Because the squares are all on top of each other, the first thing a child will learn is that they are all the same size. As they play, they will see that number 2 and number 4 both have two pieces. Here they will learn that something can have the same area, but a different shape. These would be the pieces called halves. Squares 3, 5 and 7 have four pieces. Here again, they learn that each piece has the same area, but a different shape and is called a fourth. Soon after, they will see that two pieces from square 3 or 7 would be the equivalent of one piece of square two. The same will happen with squares 4 and 5, etc.

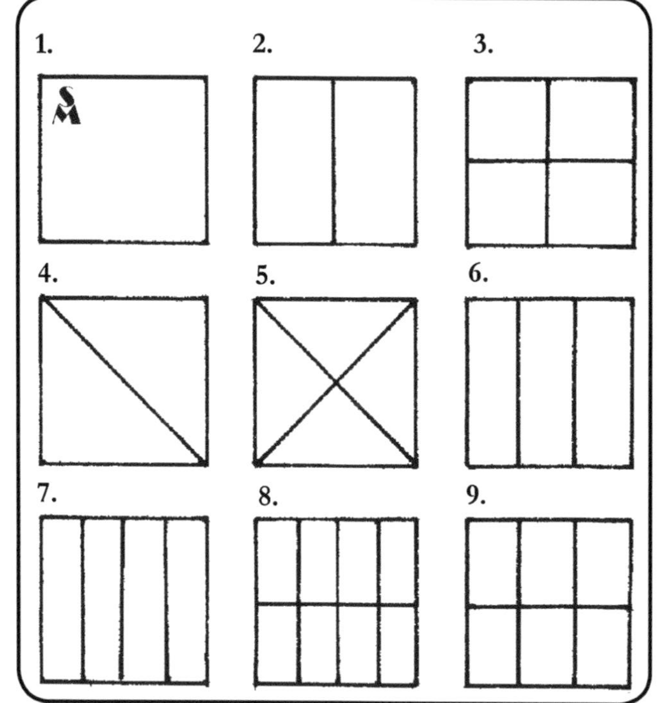

Without knowing numbers, a child will learn concepts. Early steps for adding are hidden in this toy (2+2, 3+3, 4+4, ½ + 2/4), and ultimately also multiplication and fractions (1 x 4 =4, 2 x 4 = 8, 4 x ¼ = 1), etc. Having these concepts, a child will know exactly what the teacher is teaching in grade one and later in geometry and multiplication. Children can see it in their minds and feel it in their hands. This is getting them off to a good start.

More Puzzles to Teach Math Concepts

A. 3 Boxes
Geometrical shapes,
3 layer math puzzle

B. 1 Box
Geometrical shapes,
5 layer math puzzle

C. 1 Box
12 squares,
3 layer math puzzle

D. 1 Box
All colors and shapes,
3 layer math puzzle

E. 1 Box
12 square, any shape,
3 layer math puzzle

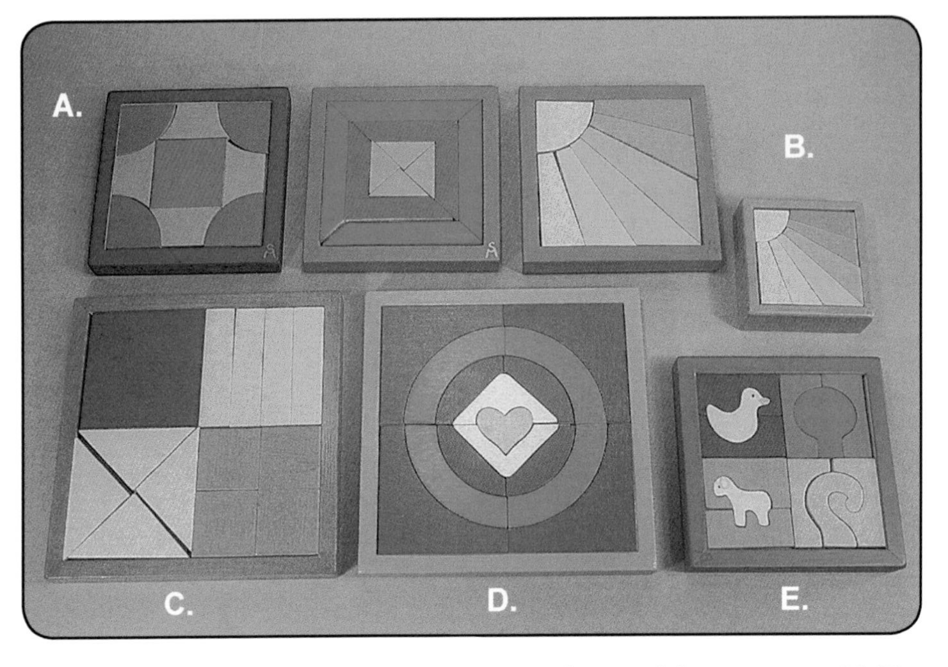

As you can see, once you know how to make boxes, you can make anything you would like to. Three, five, or nine layer boxes, and any size you want. I will not include patterns for these puzzles because you can easily design these on your own. You will be even more pleased if you do it yourself. The puzzle I designed to have all colors and shapes, cannot only be used as a puzzle, but also, the pieces can be used to make other pictures, and the children can arrange them creatively in their own free style. It is always good for children to use their own creativity. See picture on the next page.

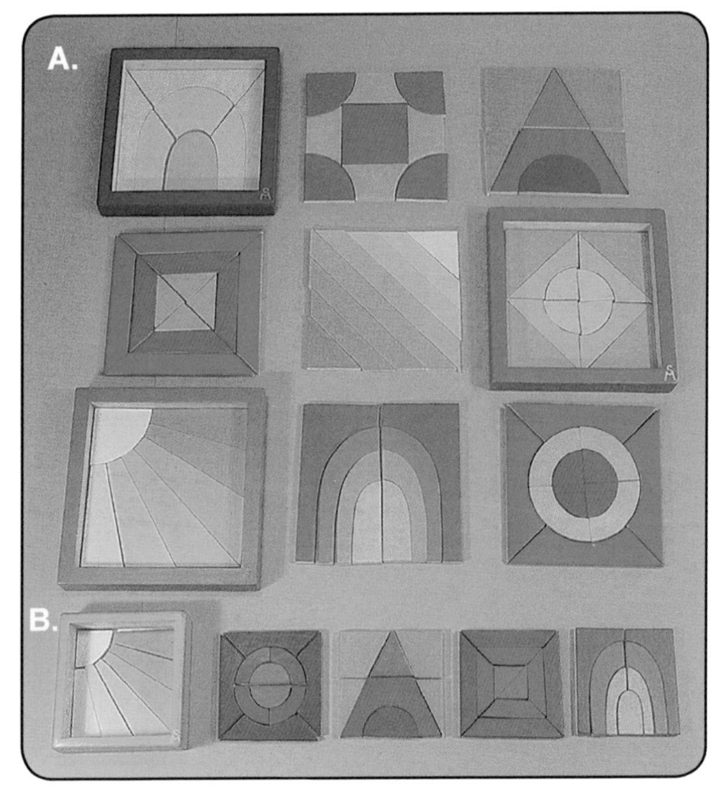

These are all puzzles that are connected to math. You can cut up a square in hundreds of different ways. You can take a paper and sketch some yourself.

Cutting straight or cutting out a circle is a bit of a challenge. By cutting animals or trees it will not matter if you fall out of line a bit. It is good to practice straight lines or circles first. The smaller bumps will of course come out when you sand. Make sure you always have a little bit of space between the puzzle and the box, never make it too tight.

When you cut out your math puzzles, you have to be very exact. If, for example, 4, 1/4 pieces are not the same size, the children will not be able to fit them into the box unless they find the exact way that they were cut.

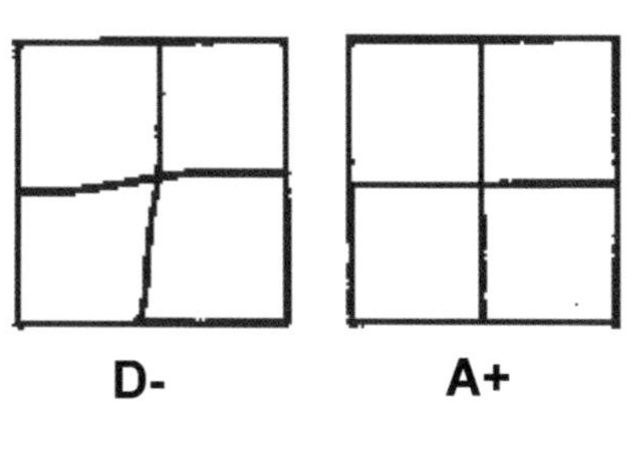

Children learn more easily by active manipulation than with just pictures in a book or on a screen.

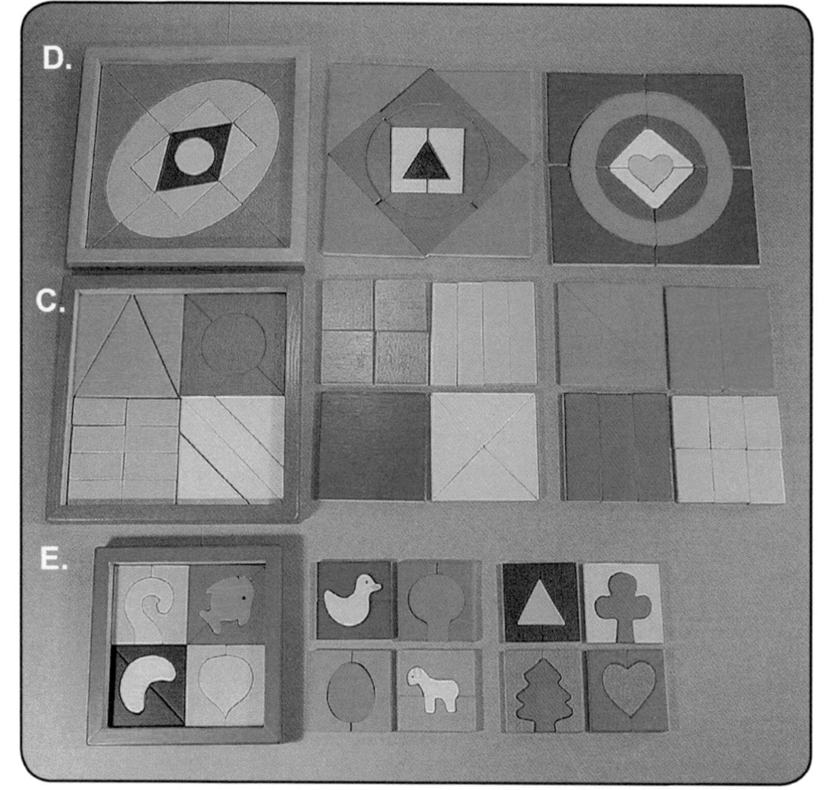

'Concepts' Children Learn as they Play

1. Counting
2. Sorting and sizing
3. Patterning
4. Volume

These different activities and learning steps interact with each other closely. For example:

- You find the same shapes in sorting as in sizing.
- Counting is a prestep to volume.
- Patterning connects counting, sorting and colors, etc.

1. Counting

Counting is one of the fundamental building blocks in math. We want to make sure that the children grasp the counting concept properly. We feel very proud when our children can rattle off numbers, but let's make sure that they understand first.

With hands-on toys, children don't just say, 1,2,3,4, they see 1,2,3,4. There is a huge difference there. They don't just see that four comes after three, they see that one number is bigger than another. They see what five fingers are, but they only really grasp the concept when they can compare with other materials in a repetitive fashion (five blocks, five children, five flowers). They seem to grasp quickly between one cookie and five!!!!! As soon as something gets interesting, the learning is intensified. By nature, children love the feeling of "I got it!" However, if life around them is boring and without stimulation, they become passive and disinterested. It is really up to us to keep their minds stimulated. By dividing things, children can learn about even and odd numbers. They will eventually learn that numbers give us a lot of information (how many children we have, how many boys and girls, what day of the month it is or what hour of the day).

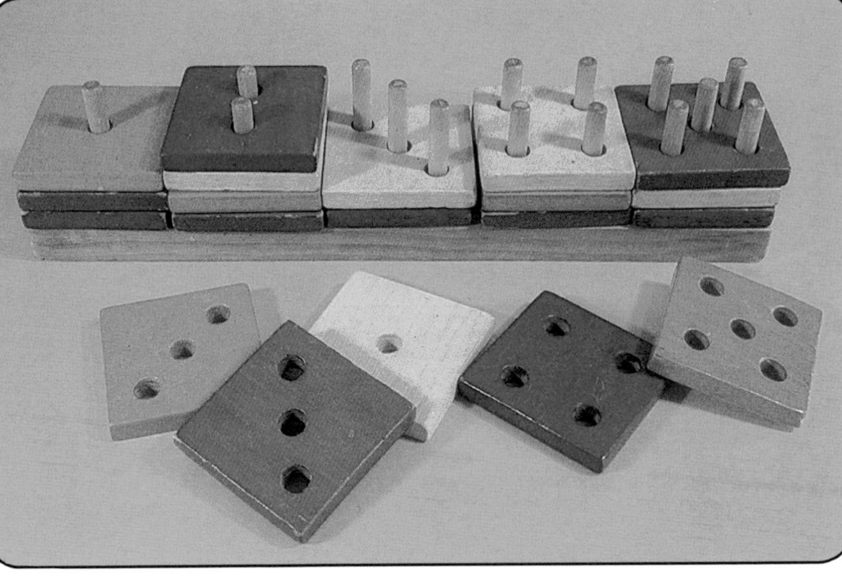

When we go shopping, we can involve the children in math by counting things into a bag. We can count one apple for each family member, or two apricots for ever person. We can also ask them how many grapefruits we need for the family, if every person gets half a grapefruit. Every once in a while when you go shopping, you will see a parent doing a great job of involving the children. Other times, parents are in a mad rush, the children are demanding and are getting shouted at and you get the feeling that it is a drag for the parents to have to take their children shopping. If you make the children feel unwanted it gives them license to misbehave.

Toys: All the toys in this math section can be used as counting toys.

2. Sorting and Sizing

Sorting is an important step or process in math. To solve math problems, we often have to first sort different components into groups before we can add or multiply. We can do simple sorting exercises for the children where they have to sort items according to color, theme or size. If they sort by theme, they first have to think what the item is used for before they can put it into a group. More complicated exercises can include two steps of sorting. With my example, I will start with easier ones and move to more difficult ones.

1. Shape and size all the same but different colors. You can have, yellow, red green and blue round beads that can be separated into containers by color.
2. Shapes different, but colors the same. If you have red beads that are square, oval and round, the children will have to separate by shape.
3. Size different but colors & shape the same. Red beads in different sizes, all round, but big, middle & small. Children often think that three big things are more in number than three little things. If you have ten of each and let the children make rows, they will see that a given number is always the same.

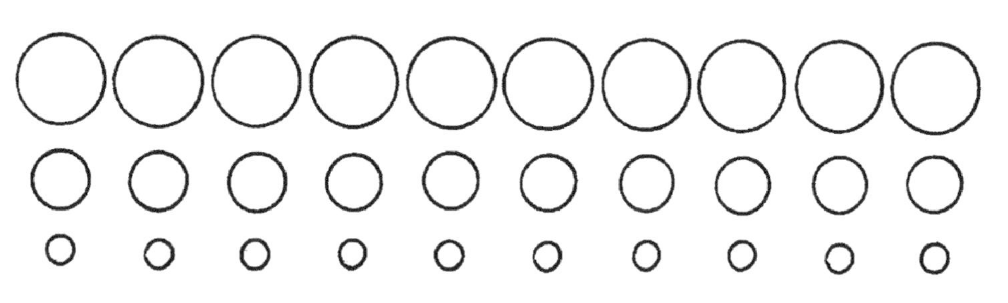

4. Shapes different and colors different. Have red, yellow, green and blue beads that are round, oval and square. You might want to separate colors and then the shapes.
5. Shapes, colors and sizes different. Have red, yellow, green and blue beads that are round, oval and square in small, medium and large sizes.
   Here too, they will first have to separate colors, then the shapes and lastly the sizes.

Cut down some milk cartons and staple them together.

Sorting often needs a plan. First, we have to think of how to tackle a problem and solve each step in the right order. I have taken beads as an example, but you can take buttons, rocks, nuts, cut out hearts, squares and circles all in different sizes and materials etc. Be inventive and use things that you have at home.

Mix two or three groups of things together and let the children think into which category they belong. Have about ten items in a group. It stimulates logic thinking that is so important for math.

Do not spend much money on these things, they are only exercises that children do for a short time. This is why it is great to use things that you have around the house. Often, I would have such an exercise ready for the children when they woke up from their nap. I would let them look at things and just let them see what they could do for themselves. Often Marc would show Nisha or I would ask for help to separate some of the mixed up stuff. It is amazing the many things that you can do to keep the children busy and help them to learn at the same time

Sorting by theme can be fun too. If you have different types of animals, you can sort them into groups; such as wild animals, animals that fly, farm, zoo and water animals. You can collect other items and let the children put them into categories. For example:

Office supplies: scissors, stapler, calculator etc.
Kitchen things: spoon, can opener, spices etc.
Play things: car, doll, crayons etc.
Forest things: sticks, moss, bark etc.
Ocean things: shells, sand, starfish etc.
Garden things: pack of seeds, soil, hose parts etc.

3. Patterning

Patterning is another math step that can be acquired though play with the right toys. You can make a toy where you have a grid-like base with little pegs to make pictures or rows in different colors.

- Children love to make patterns. For example, when they make a necklace with beads they will often use patterns such as two green, two yellow, two red, two green, two yellow and two red etc. With a peg board they can do the same thing. They can stagger lines so that they are diagonal rather than the normal horizontal. It needs a lot of concentration and children have to count to get it right.
- Math has patterns that reoccur, rules and specific steps that you have to apply. The peg board conditions the mind to think in this way and to learn concentration.
- Patterning is a pre-exercise to printing numbers in the right spots and rows for adding.
- Working horizontally and vertically is a great exercise to grasp the steps for charts.
- My peg board has room for 100 pegs (ten rows of ten pegs). Here children can easily see what is happening in multiplication. If they fill half the board, they get 5 x 10 pegs and if they divide the board into four parts, they get 5 x 5 pegs.
- Patterning is a great way to learn estimating 3 x 5 fills only a small corner, while 9 x 6 fills more than a half.
- Children will see the relationship in the length of the different amounts of pegs. When you have 8 you need 2 more, when you have 6 you need 4 more to complete a line.
- They will see that on a square, each side has the same amount of pegs.
- A peg box is also a prestep to understanding volume. You have to understand "area" before you can understand "volume".
- Children also practice perseverance and patience with this toy.
- They can sort colors.
- The perfectionist will love this toy. They feel pride and joy when the rows are completed in a straight and beautiful pattern. For other children, it is good practice in concentration, in sitting still and completing a task.

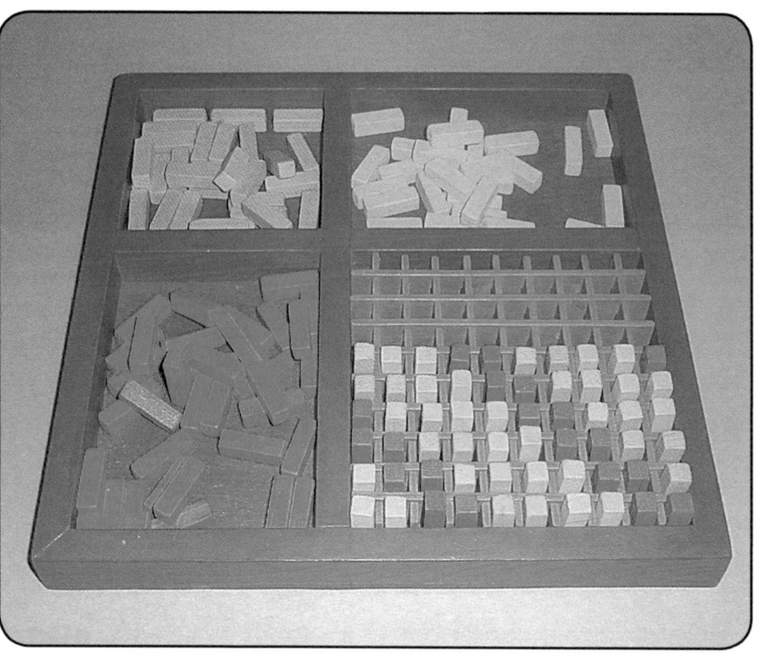

Easier version to make!

Drill holes into a board and cut dowel into pieces for pegs.

4. Volume

Volume is a three dimensional concept for math that is most definitely easier to acquire with the right toys. Here we not only have length and width, but we also have height. It includes space, not just surface or area. The decimal cube would be a classic toy to learn volume. In it, you have ten trays of ten by ten little blocks or spaces that make 1000. For small children, you only want to teach the concept of space, which is simple and a precursor to the decimal cube and the basics on how to square and cube a number. I made a little box in which two cubes of exactly the same size fit. One of the cubes is divided into eight medium cubes and one of these eight cubes is again divided into eight smaller cubes. In total, there is one large cube, seven medium cubes and eight little cubes. Children will realize, that as soon as you divide space (height and area) you get many more pieces than if you just divide a surface. They will picture the box full of medium cubes or even full of little cubes. For medium cubes you would need 16, and for the small ones, you would need 128 to fill a box. If you have a lot of cubes, children can just build with those and learn many things as they use them in a variety of ways. For example: (See photo on next page)

Length: If you have two blocks and want to make it longer by **one block**, you only need to add one block 2+1 = 3.

Area: If you have two blocks squared (2 x 2 = 4), to make it bigger by **one block**, you need to add five blocks (3 x 3 = 9).

Volume: If you have a cube of two blocks wide and two blocks high (2 x 2 x 2 = 8) to make it bigger by **one block** you need to add 19 blocks (3 x 3 x 3 = 27).

You can see, as children play with blocks, they learn about volume. They don't say, "I am going to cube this number by three," they just say, "I want to make this house bigger." They will be surprised how many blocks that they need. Sets of blocks are great to have. Children pick up many concepts as they build. Blocks are a great teaching tool for math, either the little table block sets or the large floor sets. (Ikea furniture store has some nice sets).

More Math Toys for Learning Volume Concepts

1. DISC PYRAMID - 2. CUBE BOXES - 3. STICK PUZZLES

4. 3D PUZZLES - 5. VILLAGE MADE WITH BLOCKS

1. Disc Pyramid: Cut out wood circles in different sizes. Put on top of each other and determine how long your dowel has to be. Make the dowel a little longer than the stack so that the children can hold onto it. Drill a hole into the bottom disc and glue in the dowel. Drill holes into all the other circles slightly larger so that they fit easily over the dowel. When you use your protractor, make sure you mark the middle well, so that you know where the holes should be. For this project, you can take ½" (12 mm) or 1/4" (6 mm) plywood and choose your own color scheme. You can have more discs or less discs depending on the age of the child you are making it for. I have taken ½" (12 mm) plywood. This toy would be suitable for a child about 1 ½ to 2 ½ years old.

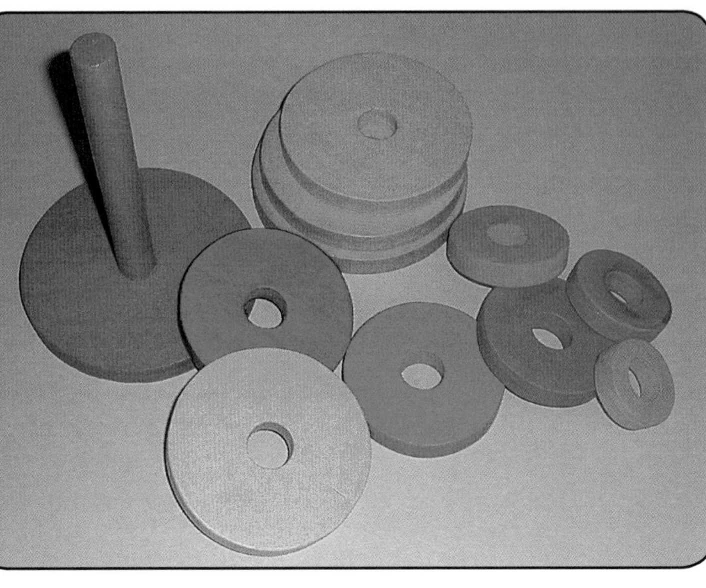

2. Cube Boxes: I bought the cubes in a lumber store and just made a box for them. You can paint them with six different colors (a cube has six sides) so that children can make patterns. You can leave them just plain, especially if you want to do math exercises. If you make a box with 10 by 10 cubes, it would be great to practice multiplication tables because a child can see exactly what happens. If you sit down with a child to play with such a box when they are going through learning the multiplication table at school, it will be very rewarding but will also demand a lot of patience. To build confidence and to encourage is very important. How can we grasp something if we think that we cannot do it or someone tells us, "you will never get it!"

3. Stick Puzzles: Here too I bought ready cut sticks and just cut them to the desired length. You can determine the size yourself. One box is a four seasons apple tree and the other box just has different pictures on it; a duck, a heart, a tree, a fish etc. With this toy, you can make it either easier or harder for the children to do. I remember as a child having to put together a box with cubes and pictures of different fairy tales on them. We soon figured out that we could take a whole row of cubes and flip them over to get the next picture. The same can happen with the stick puzzle. If you paint one side of your whole stick puzzle and then turn it over for the next picture, the children will catch on quickly. However, if you turn them over and move them around, the children will really have to find the right part for the next picture.

4. Three-Layered Puzzles with three different thicknesses of Wood: If you mix different thicknesses of wood, you gain a special effect of depth. At the same time, you are adding something new to the puzzle, so the child will learn a new concept and show new interest. I have three toys that I have made in this way:

A. House with building blocks.

B. Heart in a tin.

C. Bears in a tin.

A. House with Building Blocks: I cut out three houses, each from 1/8", 1/4", ½" (3, 6, 12 mm) plywood. I then cut these houses up into blocks for the walls and triangles for the roof. All the pieces are the same size so that they can be interchanged. I painted each house a sepa-

rate color; the 1/8" house with yellow, the 1/4" house with light orange and the ½" house with dark orange. Choose colors that go well together, so that when you mix the building blocks they blend well. You can build little Mediterranean villages, castles or simply individual houses. When the child has finished playing, everything folds back into the box that has

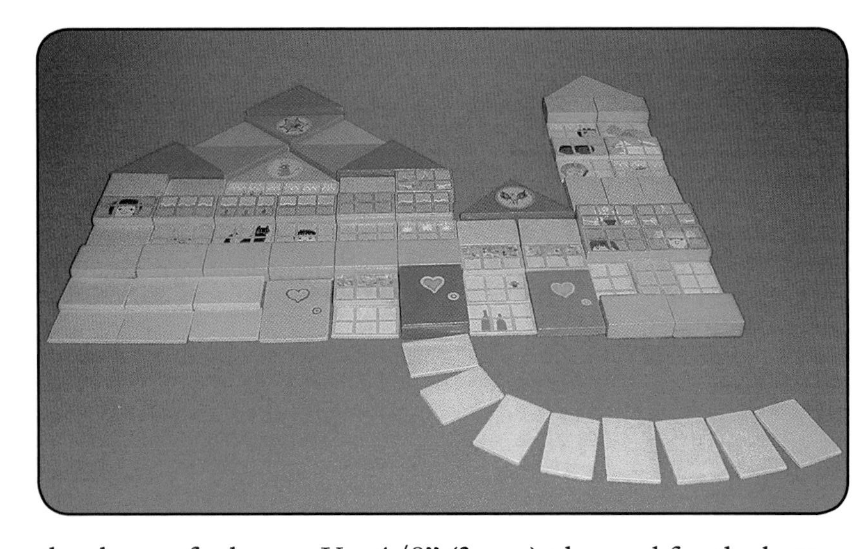

the shape of a house. Use 1/8" (3 mm) plywood for the bottom part of the box and moldings as in Page 16, Box 3. When you put the pieces back into the box, you can do it in any way you want as along as you take one thin, one medium, and one large piece to put on top of one another. Children can practice creative thinking with this puzzle.

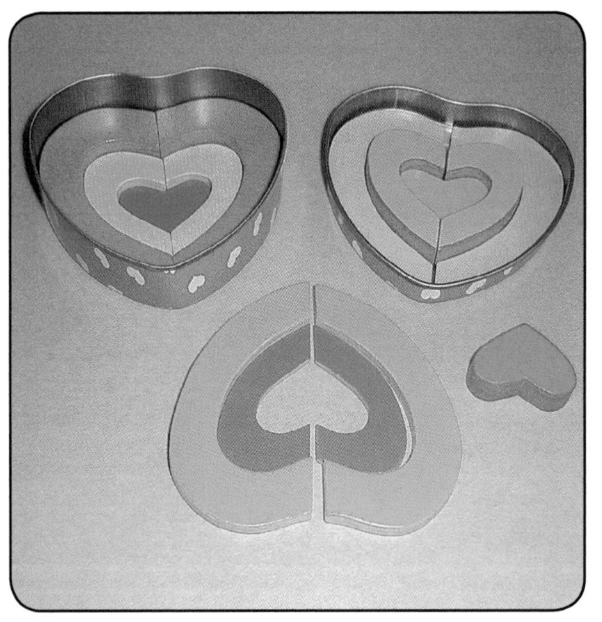

B. Heart in a Tin and Three Bears in a Tin: Sometimes you find a nice box or tin that you can use for a puzzle. I painted the heart puzzle with six colors so that when you start mixing the pieces together, you get a nice effect. Each side is a different color, but all the pieces that belong to the same side are the same color. Since the bear tin shows a picture of the bears picnic on the front, I added a blanket with a picnic basket. The wood for the bears are big, medium and small in thickness and the cups and bowls also come in big, medium and small.

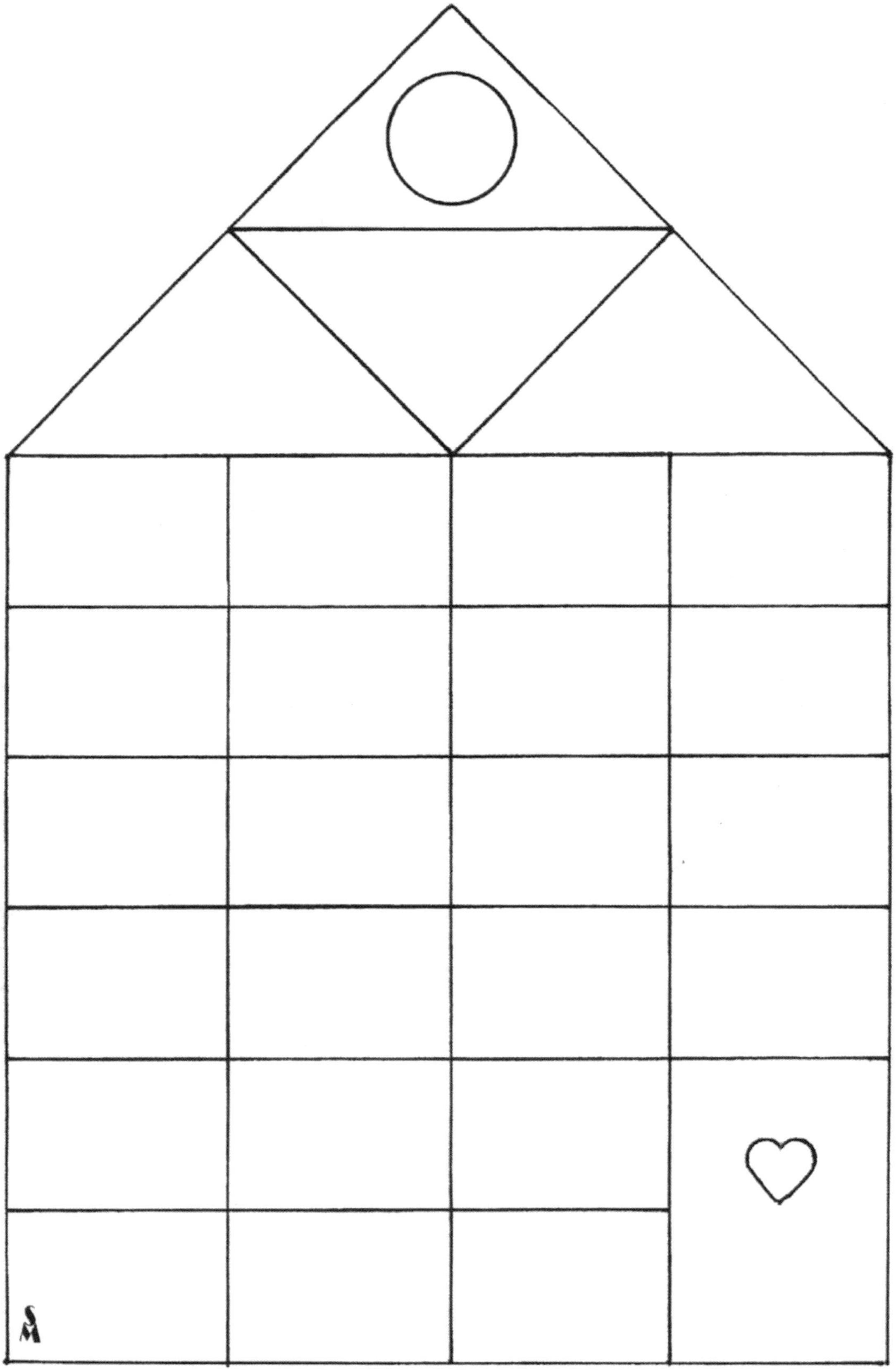

5. Village made with Blocks: Be an architect -- select shapes and colors to build houses.

This is a small tray with 14 blocks:

4 Cubes - 1 ½" x 1 ½" x 1 ½" (4cm x 4 cm x 4cm)
1 Double Cube - 3" x 1½" x 1½" (8cm x 4cm x 4cm)
2 Half Cubes - 1½" x 1½" x ¾" (4cm x 4 cm x 2cm)
1 Double Half Cube - 3" x 1½" x ¾" (8cm x 4 cm x 2 cm)
6 Triangles - 3" (long side) 8cm (long side)

Each block is painted with 4 different colors. On most of them I painted little details, some are just plain. This gives a child the chance to either use just one color at the front or mix and match. With a few hints you can help them be more creative.

When you look at the photos you'll see that you can come up with many more variations!

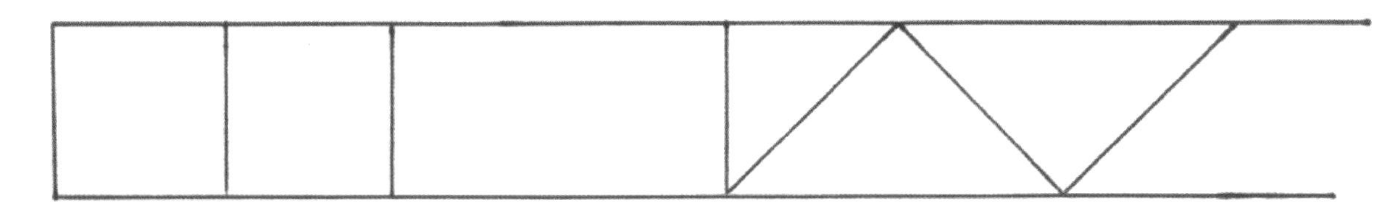

I bought ready-made sticks at the lumber store:

One length of 1½" x 1½" (4cm x 4cm) and
One length of 1½" x ¾" (4cm x 2cm)

You can design your own toy with more blocks, different colors and more shapes. You choose and build on one side and then see what it looks like on the other side. Surprise!

Additional Ways of Learning Math Concepts

1. Hammering game.

2. The little felt board.

3. The cookie sheet with magnetic strips and buttons.

These types of toys have the same learning capacity for grasping math concepts, but they are different in materials and activity. What a child learns with one, will be enforced with another. They all have geometrical pieces in different sizes and colors and can be made into pictures or mosaics. With these toys, children learn counting, sorting and patterning.

1. The Hammering Game

Size: Approximately 15" by 10" (38 x 26 cm) with the frame.

Box: I bought some molding strips in the lumber store, cut them to size and used glue and small nails to fix onto the plywood. You can also use heavy cardboard for the base. Inside, the piece is made of a soft press board that is usually used as a ceiling tile. This can easily be cut with an exacto knife (See Box 3, page 102).

Pieces: I cut plywood scraps into different shapes and drilled small holes, a little larger than the thickness of the nails. In larger pieces, I would drill two holes.

Nails: Find nails that have large heads and have a length that is a little shorter than the thickness of the press board and the plywood pieces together. I never put many nails in a box at a time.

Hammer: For safety reasons, find a really small hammer or make one from pieces of hardwood dowel.

• **For the Home:** You will know when your child is old enough and does not put the nails in his/her mouth.

• **For a Childcare Center:** Do not start too early and be sure to supervise. As you can see in the picture, Nisha is about 2 ½ years old and is just nailing pieces on in random order, mainly enjoying the action and mastering some small motor skills. Marc, on the other hand is about 3 ½ years old and is working with a plan and a purpose. The same toy can offer different learning experiences for different ages. This is why you can bring out the same toy over and over, and the child will always make new discoveries.

2. The Little Felt Board: You can easily put together this toy in one evening and it will cost hardly anything. You need: One cardboard for the base, two felt squares; one to glue onto the cardboard and the other to make a frame. After gluing it, press it under a stack of books so that it does not bend. Use your felt scraps and cut up triangles, circles, squares, strips etc. with all different colors. Children can put the pieces on top of each other in order to add a visual effect of depth. The expensive part is, that you buy a nice box of chocolates so that you can store the felt pieces in a flat box for easy access.

3. The Cookie Sheet with Magnetic Strips: You can either use a plain cookie sheet or you can paint a simple landscape on it. Cover it with clear mactac for protection against scratching. Look for magnetic strips, buttons (or other shapes) in office, hardware and toy stores. I found little squares the other day which really got me excited! Store pieces in a cookie tin, so that the magnets will keep their strength. Both sides of the cookie sheet can be used.

You could also cut out different wooden pieces as in the hammering game and put a piece of magnetic strip on the back to use on the cookie sheet. They could also be made from cardboard or Fun Foam instead of wood. For either of these toys, do not be tempted to cut out little figures or trees. The idea is for the child to think of how to make a house or a tree themselves instead of just taking the finished one out of the box.

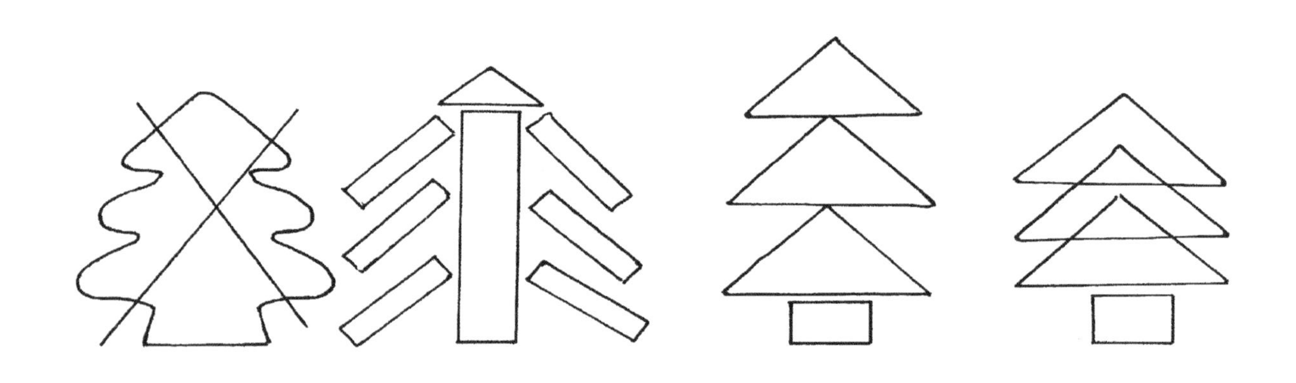

SCIENCE PUZZLES

1. The Frog
2. The Duck
3. The Butterfly
4. The Apple
5. The Four Eggs
6. Twelve Animals
7. Four Seasons Apple Tree
8. The Fish Puzzle

Education is:

Watching a child being a work of art in progress!
It is like seeing a masterpiece unfold and completed.

The following Science Puzzles have all the same squares:

1. **The Frog Puzzle**
2. **The Duck Puzzle**
3. **The Butterfly Puzzle**
4. **The Apple Puzzle**

(4 Small Science Boxes
3 squares, 3 layers each)

5. **4 Eggs Puzzle**
6. **12 Animals Puzzle**

(2 Large Science Boxes
12 squares, 3 layers each)

Precut all squares from ¼" (6 mm) plywood. I have put all these puzzles into the same section because they are all made with the same size of squares (3", 7.6 cm). For the small science puzzle boxes, you can use the same moldings as for the larger puzzles, where I use the molding shown on Page 16, Box 3. I have cut down the molding for the small puzzles as shown in Box 2. To make these puzzles you can use the same steps as I explained in the previous pages of this chapter. The four eggs puzzle is quite intricate, so you need a bit of experience to cut it out. You have to also be careful that you do not break off the corners when you are sanding the wood. For the 12 animal puzzle and the four-egg puzzle, I have cut each square into three pieces. I left the animal whole and cut the background into two pieces. If the animal stays whole then the children can still play with them.

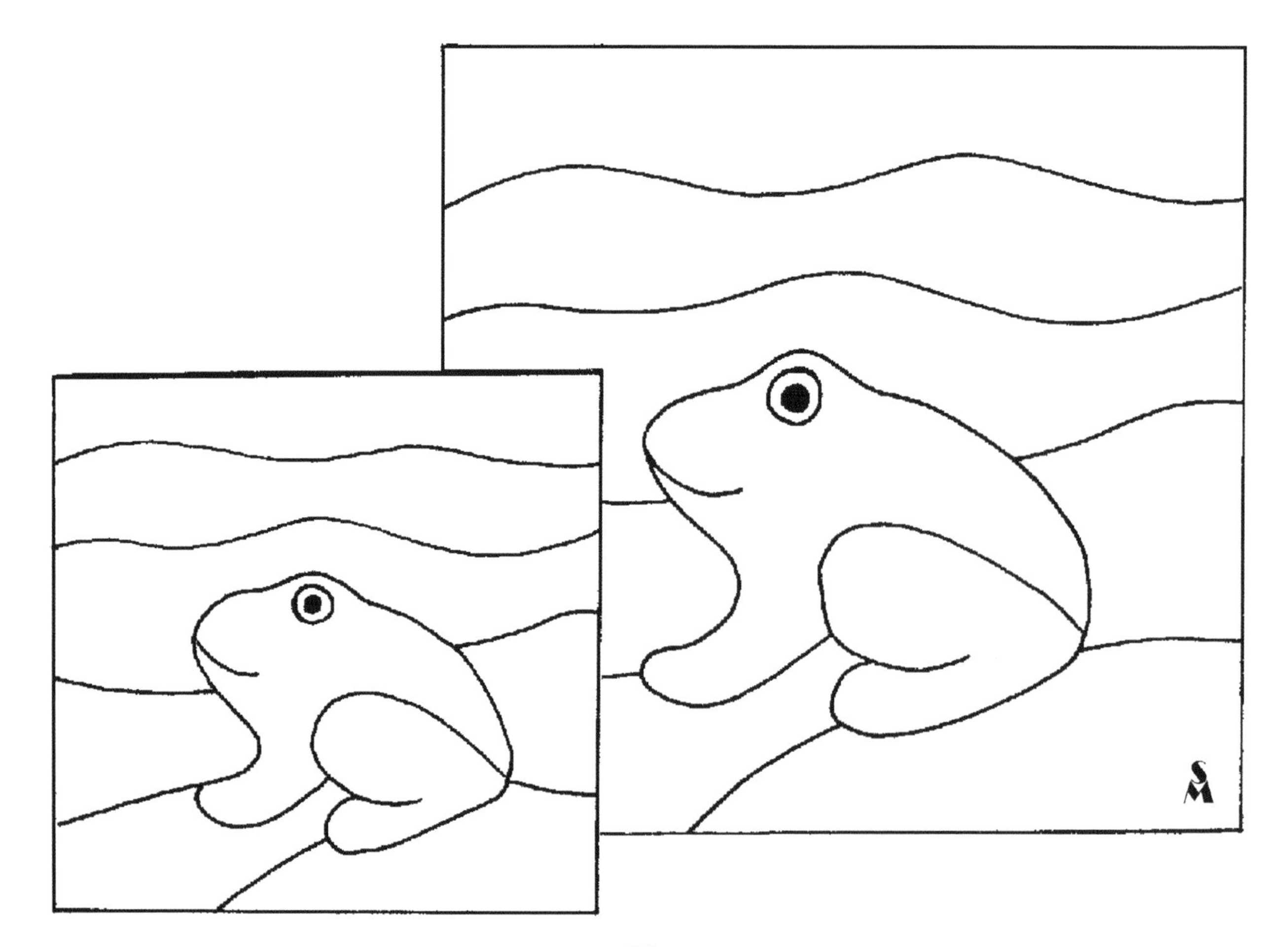

I cut apart the egg so that it looks as though the animal has just hatched. Do not make very small pieces, since they will only get lost. Leave the butterfly egg attached to the branch. I also drilled a hole into the cocoon so that the children can see that the caterpillar has hatched. For the background pieces, use only one color, otherwise it will get confusing. Your main figure will stand out really well if you use just one color, except if you cut the pieces to make a special background as in the Frog puzzle. I have not given you exact dimensions because I do not know what types of moldings you will find. The pattern of the puzzle will have the right dimensions. All you have to do is take the square, add 1/8" (3 mm) extra space and then see how big the box has to be. You can enlarge the puzzle on a copying machine, if you like bigger pieces. This can make it easier for smaller children. See the Frog pattern as an example. (Page 36).

With these science puzzles I wanted to make the children more aware of different cycles in nature. There is so much to observe, learn and enjoy by studying nature. It is a very wholesome activity and teaches many facts about new life, transformation, patience, etc. The little science puzzles are as follows:

Egg	Caterpillar	Butterfly
Egg	Tadpole	Frog
Flower	Apple	Seed
Egg	Little Duck	Big Duck

The four egg puzzle has different types of animals that all hatch from eggs. I have made four different sizes of eggs so children can see which animal belongs to which egg. Smaller children might only see a difference between the biggest and smallest egg and the ones in between would be more difficult to place. Then there is the perfectionist who wants to fit all the animals perfectly, or the more free spirited child would love to put the penquin into the dinosaur egg. This puzzle has one layer of eggs and two layers of animals.

Large Egg	Ostrich	Dinosaur
Medium Large Egg	Aligator	Swan
Medium Small Egg	Fish	Turtle
Small Egg	Penguin	Duck

For a little fun I added a baby picture of my children to the bottom of the box. This way they could fit themselves into an egg! It gave us the opportunity to talk about my pregnancy and how they grew in my tummy.

7. The Four Seasons Apple Tree

Science Puzzle, 3 layer, 1/4" (6 mm) plywood

Precut:

* One piece of 1/8" (3 mm) plywood
* Three pieces of 1/4" (6 mm) plywood
* Four pieces of molding as in Box 2, Page 16. Cut down to fit the four sheets of plywood. Frame can be a little higher

This puzzle may seem to be more complicated, but by studying the photos and the pattern, you will see how it is done. The molding is glued onto the puzzle as a frame to finish it off nicely. This time the molding is not part of the box. I have only cut out the trees and left the background whole. Because you would never be able to cut out each tree the same, it would never fit if you make them the same size. Due to this, I cut them out in different sizes. The winter with the bare tree is painted on the bottom piece (1/8", 3 mm plywood) and cannot be taken apart. The fall is the smallest tree, the summer medium and spring the largest. Since the spring tree is on top, I have made a drill start so that it looks neat. With the other two trees, I have started by cutting out right from the edge down by the trunk. It will be glued together so that you won't see the cut.

Once you have cut out the three trees, you only have to sand the insides really well and then glue the background pieces of the trees (starting with the smallest one) together onto the 1/8" (3 mm) piece. Make sure you always know which is the right side up of the pieces. Once the four pieces of plywood are glued together, you can sand the outside and glue the molding frame on, holding it together with elastic. When it is dry, you can sand the corners and the edges. I do not use any nails for the frame because glue seems to be enough. You can now cut up your trees. I cut the trees into seven pieces, but you can make double the amount of pieces if it is meant for a child who is 4 or 5 years old. Sand all the pieces well and your tree is ready to paint. I do not paint the sides of the pieces since you may then run into some space problems. You would have to sand quite a bit more to give space to the paint and the varnish.

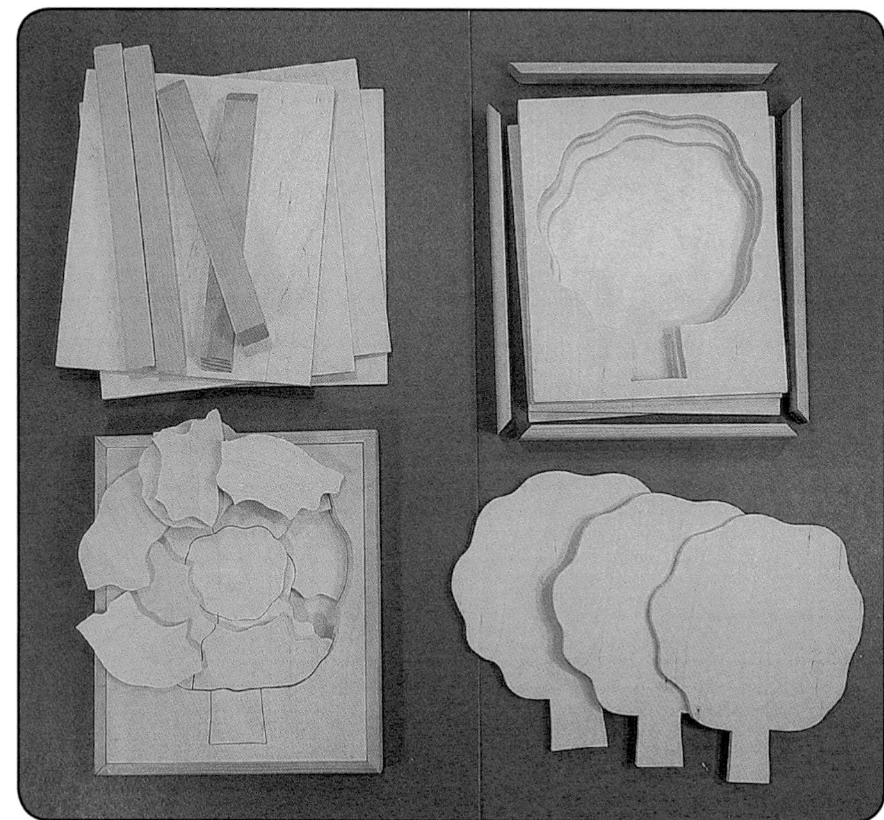

Trace these three different sized trees onto the 3 pieces of 1/4" (6 mm) plywood. Cut up the pieces of each tree differently so the children don't try to jam the pieces in where they don't belong.

(Note: the original patttern is 15% larger.)

8. The Fish Puzzle

Science Puzzle, 1 layer, 1/4” (6 mm) plywood

* Add 3/4” (2 cm) for a frame all around, to the pattern
* Precut two pieces of plywood, one 1/4” (6 mm) and one 1/8” (3 mm).

Trace the pattern inclusive frame onto 1/4” sheet. Make a drill start along the frame and complete the puzzle as shown earlier in this chapter.

This is a very intricate puzzle, where every piece is a sea creature. I have painted them in bright colors starting with yellow tones in one corner and working my way through oranges, reds, pinks, purples and blues. One fish and the frame are green. With this puzzle a child can play “ocean.” You can lay out a blue fabric sheet or a blue poster board for an ocean. Children will sort fishes by color, size or by their friendly or fierce faces. It seems to be a fairly difficult puzzle to put together, but once you see that you can start with one color and add another color as you go, it is not as difficult.

(Note: Enlarge pattern 20%)

At this point I would like to give you more encouragement

Aristotle said,
"What we have to learn to do, we learn by doing".

Or we say,
"Nobody is born an expert".

Each project has a number of steps. First you prepare the material. Once you have the material in your hands it is easy to figure out your first step and then your next one, etc. It is amazing to see how everybody can do the project once they put their mind to it. It might not be perfect in the beginning or you might make a mistake but this is how we learn. If you never try you don't have the right to say that you can't do it. I have seen beautiful puppets made by people who first said that they couldn't sew! Usually the lopsided one is the one with character and ends up being the children's favourite. They know how to appreciate something homemade. Even if you try to copy something it always is a little different and has your personal touch.

The joy of giving children a toy that you have made yourself is worth the effort. Quietly watching them as they play with it and realizing that they are in the process of learning is priceless. These are the little things that make child rearing special. We need moments such as these in the long haul during the years when the children are little. At the time at least it seems long. However I can assure you that when you look back to the years when the children were little it seems very short and you regret that you didn't enjoy it more. Remember that these opportunities will not come back. The real joy is that you can watch your children grow up and become well-rounded citizens. You know that you are the one that helped them get there with your love and effort. Nothing compares with the feeling of a job well done. I came to realize that I could only be fulfilled by fulfilling the job that was given me, as my vocation, as a spouse and parent. However you can expect to make some mistakes along the way.

Even if you have to work there are ways to make child rearing a priority.

Magnetic Toys for Math and Science

I have already introduced the iron rock, the cookie sheet, the cookie tin and the magnetic stage in Book 1, "Educational Storytelling," but there are many other ways to create magnetic toys. Once you start with a few ideas, more will come. You can start your own collection with home made ones, and with materials that you already own. Everyone has a few cookie tins and if not, they can be found second hand. It would be nice for you to add an iron rock to your magnetic box.

I always start my presentation on magnetics with an iron rock. It stimulates thinking and raises many questions. To be able to pick up an ordinary rock with iron in it using a magnet, surprises us. The iron content is so high that it is like a lump of iron. We compare other rocks that the magnet does not pick up and explore different types of metals. We can compare the iron rock to cast iron or to steel and see how a process can change material. We can talk about the magnetism and gravity of our earth and how we would all fly away like astronauts if the earth wasn't the biggest magnet we know. We can talk about qualities and workmanship of metals or show the children a picture of a large smelting pot where iron is melting. We can also talk about miners and how they bring iron out of the mountains.

After the introduction we can let the children find iron or steel in the home or the classroom. You can even hide some iron objects in the garden and let the children find them with magnets. Children love this game. Magnetic toys are very suitable in many learning games. You can teach colors, shapes, sorting, language, singing, small motor skills, concentration, creative thinking and much more.

Magnetic Toys in this Section

1. Experimenting Box
2. See-through Boxes
3. Finger Puppet Stand and Magnetic Rings
4. Fishing Game
5. Fruit Tree
6. Magnetic Faces
7. Butterfly
8. Matching Game
9. Color Matching
10. Cookie Sheets

1. Experimenting Box:

Collect a variety of pieces of different metals; nails, washers, bottle caps, bolts, lids, buttons etc. Make sure they are all large enough so that they cannot be swallowed. Have some brass, aluminum and copper pennies mixed in with the iron. Next add a magnet and just let the children explore. Large magnets are fun and can often be found in farm stores. Farmers use these to take out pieces of iron that a cow may have ingested along with the grass.

2. See-through Box:

You can make a variety of boxes with different items inside. I have taken some old picture frames that have depth to them, painted them and then cut pieces of Plexiglas to replace the glass. You don't want to use glass for safety reasons. I glued the Plexiglas inside, put pieces of iron in each box and sealed it tight with a piece of cardboard on the bottom. Some children are very curious about how to open the box!!!

*As you can see, I put small nails into one of my boxes. You cannot let the children work with small nails, but if they are contained like this, it is OK.

*One box has about 25, 3/8" steel balls from a wheel bearing. Ask a car mechanic for an old one. Degrease them first.

*The most fun of all is the iron filings. If you know someone who works in a metal shop, they will gladly give you some. Another great thing you can do is go to the beach and pass the magnet through the sand. If you are lucky you might find magnetic sand. We found patches of black sand at the beach and the magnet picked up half of the sand. This was exciting! For this box you can have any type of magnet with fairly good strength. Move it along the bottom of the box and watch the iron pieces move around. If you use a magnetic bar, you can see the magnetic fields in the iron dust.

3. Finger Puppet Stand and Magnetic Donuts:

If you make a finger puppet stand, you can also use it for donut magnets. It is very fascinating for the children to see that on one side they attract and on the other repel and float in the air. Include a few washers.

4. Fishing Game:

Fish tank: Take four pieces of cardboard and drill holes along the sides. Sew them together with a shoelace stitch, not too tight. The idea is to be able to collapse it for easy storage. You can also take a simple box and cut out the bottom. You can paint the box, decorate it with wallpaper, stickers, put pressed seaweed on it etc. This is a great rainy day project for the whole family.

Fish: I have cut out wooden fish, drilled holes the size of small round magnets and inserted the magnets into these holes. You can make the hole where the fish's eye should be. Make the hole a little tight and gently hammer in the magnet. You can let the children paint the fish. You can even cut the fish from plastic coffee lids or from Fun Foam. Decorate the fish with a permanent marker and put a paper clip on it instead of the magnet. Plastic fish can be used in a water table, a small pool or even a bucket.

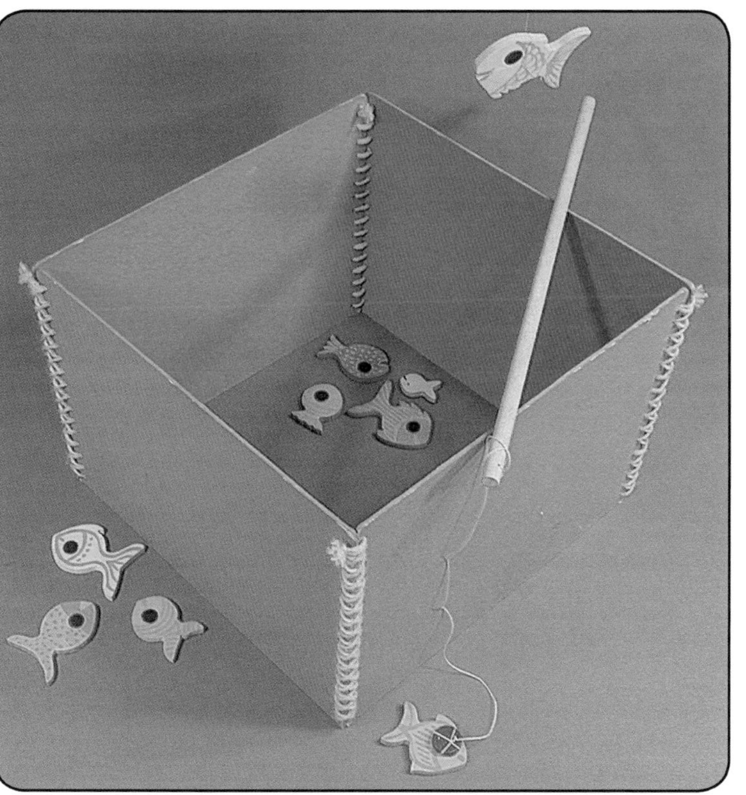

Fishing rod: Any stick or two with a string and a large magnet tied to the end. It will easily pick up the fish with the paper clip too. Magnetic rings or horseshoe magnets attach easily to a string. With this toy you can simply go fishing, or you can play games to see who can catch the biggest fish. You can write numbers on the fish and add them up, or you can teach the different colors. You can make up a game with your own rules.

5. Fruit Tree

This is one of my favorite magnetic toys that you can easily make yourself. You will need:

* Cookie tin: take a plain cookie tin, paint a simple tree with some leaves on the lid and a few green plants around the sides. Cover the painted area with clear mac-tac for protection, after the paint is completely dry.

* Baskets: find four small baskets that fit nicely inside the tin. If you have too much space, they will always fall over inside so it is a good idea to coordinate the sizes of the cookie tin and the baskets well.

* Fruits and flowers: cut these out of doubled felt. Then before you sew them together, put a little piece of leather in for a stem and a strong button magnet. I have made apples, pears and plums. Flowers can be made all the same or in a variety of colors. I will not include a pattern because this one can be done quite easily by yourself.

Add a cookie sheet or stovetop cover to the exercise for variety.

6. Magnetic Faces:

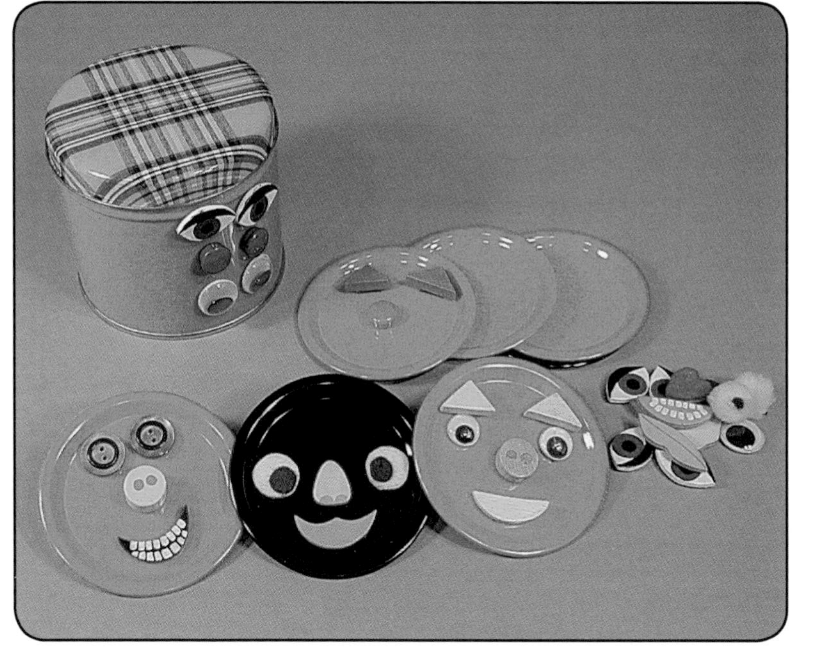

Plates and lids: I have found tin coaster plates that work well. You can take canning lids from wide mouth jars that have no bumpy writing on them. Cover them with a circle of nice plain color mactac.

Eyes, noses and mouths: Cut out from wood, cardboard, or just double glue some felt pieces. Use pompoms, googli-eyes, or buttons in different sizes (not too small or they could be swallowed). Put magnets on the back of these pieces.

Tin can: Preferably a plain tin can so that the children can line up the noses, eyes and mouths and select them as they play. It is harder for them to see if there is a pattern on the tin.

This toy can be used when you talk about people, feelings, and family. Children can make happy and sad faces and learn how an expression changes as the eyes move closer or further apart. Two children playing together can have fun showing each other the different expressions they have made. You can even include pieces of fur for hair. The sorting and counting of the pieces teach concepts of math.

7. Butterflies

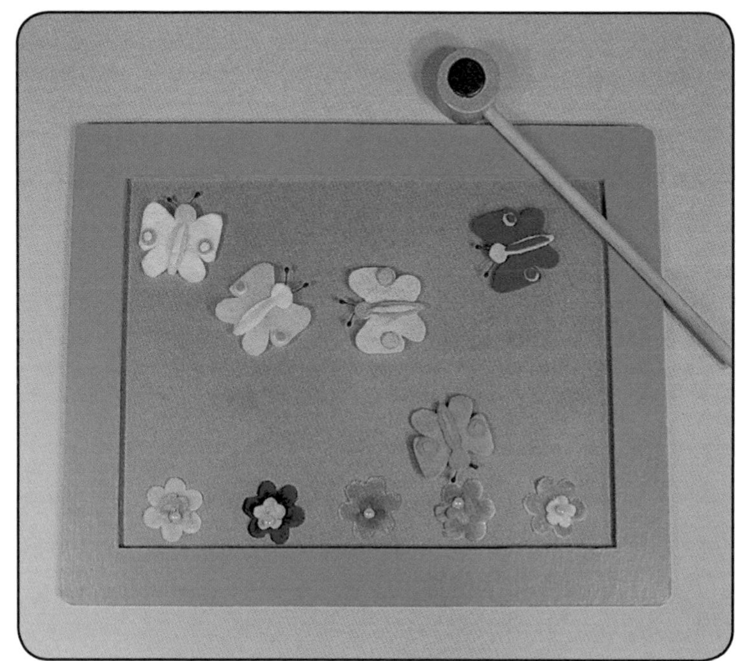

This is another matching game but with another twist. The frame is made with a very heavy cardboard and the inside is a piece of poster board, 11" by 9" (28 x 23 cm). You can move the magnetic butterflies which are made of felt, with the magnetic handle from underneath the posterboard. I have glued the flowers on so that you can only move the butterflies. In this way, children will have to move the pink butterfly to the pink flower and in the process will learn about colors and matching. I have made a little poem to go with it:

"5 little butterflies are looking for food, they like to eat honey because it tastes good.
Here are the flowers that give us our treat, lets find the right color - and then they can eat."

8. Matching Game:

Cookie tin: Again find a nice cookie tin. It is good to store the magnets in a cookie tin as they will keep their strength this way.

Figures: Cut out from 1/8" (3 mm) plywood, sand them well and paint them. You can use cardboard and felt pens as an alternative. Use adhesive, magnetic strips on the unpainted side. In this game I have included:

* Cow: bottle of milk
* Chicken: egg
* Sheep: wool
* Tree: log
* Goose: pillow
* Fruit tree: basket with apples
* Rock: iron
* Caterpillar: butterfly

You will have some more ideas that you can add.

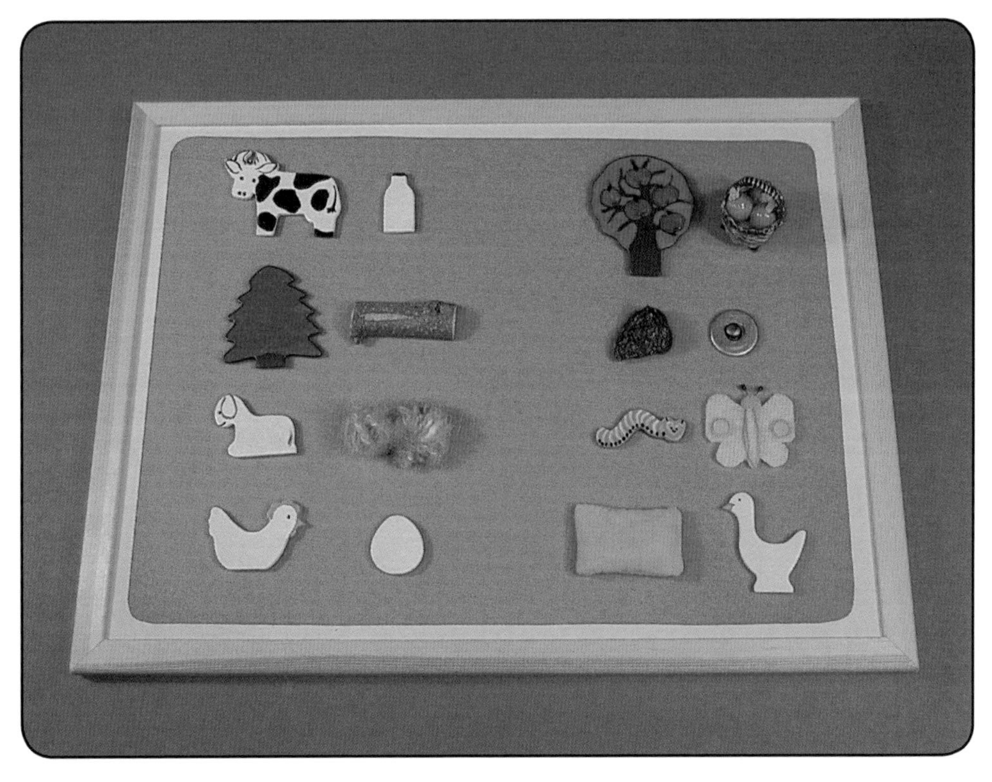

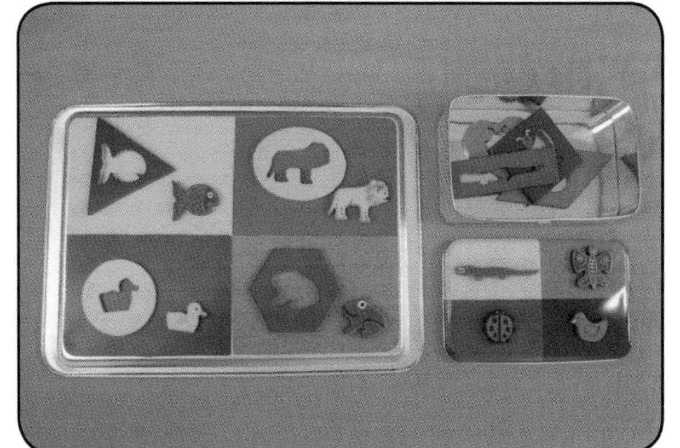

9. Color Matching

This is a mini-puzzle exercise with magnets where children can learn concepts of matching and shapes. It includes a cookie sheet and a tin can with MacTac in the four primary colors as well as eight small puzzles with magnets on the back of the animals.

You can use any puzzle and put magnets on the back, it will bring another more exciting aspect to a puzzle.

Design your own shapes and animals, or have other shapes within the shapes instead of animals. There are many different ways of making an exercise, it gives you an opportunity to realize your own ideas!

Let children explore on their own, have all the pieces in the tin can and the cookie sheet ready beside it. Watch a child approach a new toy, see how they discover by manipulating pieces, by being excited to feel competence, by being able to match and to find out by themselves. They can use their own creative arranging, learn to make decisions, feel the shapes in their hands and compare them with others. Let them show you what they have discovered, fill in words of shapes and colors casually. Watch how they progress and extend a learning process as they play with the same toy again and again over a period of days until the learning process is exhausted and they move on. Children come back to the same toy after it has been put away for a time and discover new concepts because in the meantime their understanding has grown.

10. Cookie Sheets (Tin only)

Some of the easiest and most rewarding activities you can create are with cookie sheets and tin cans. Take a magnet along when you buy them. Aluminum or stainless steel don't work and the fancy non-stick ones are not suitable because nothing sticks, it has to be tin. You can paint them very simply even with acrylic paint if you cover them with see-through MacTac also. Bright colors are nice half blue, half green, make a good background for different themes.

Study the photographs and it will give you many more ideas for matching exercises for math exercises, to sing songs, to just use your fridge magnets and put together stories.

<u>Magnets</u>: Use magnetic sheet, magnetic strips or magnetic buttons.

<u>Pieces</u>: There are many easy ways to make the pieces for different themes. Put stickers on a magnetic sheet and cut them out. Use fun foam, there are many shapes cut out already, just add magnetic sheet or strips. Use buttons, googley eyes, etc. These make nice table toys. We need more educational table toys where children can sit, think, reflect, build, solve problems and be creative. Craft stores have little wooden pieces that are suitable for all kinds of themes.

All the following cookie sheet exercises have Math and/or Science aspects to them. There is counting, sorting, shapes, weather, animal life, etc.

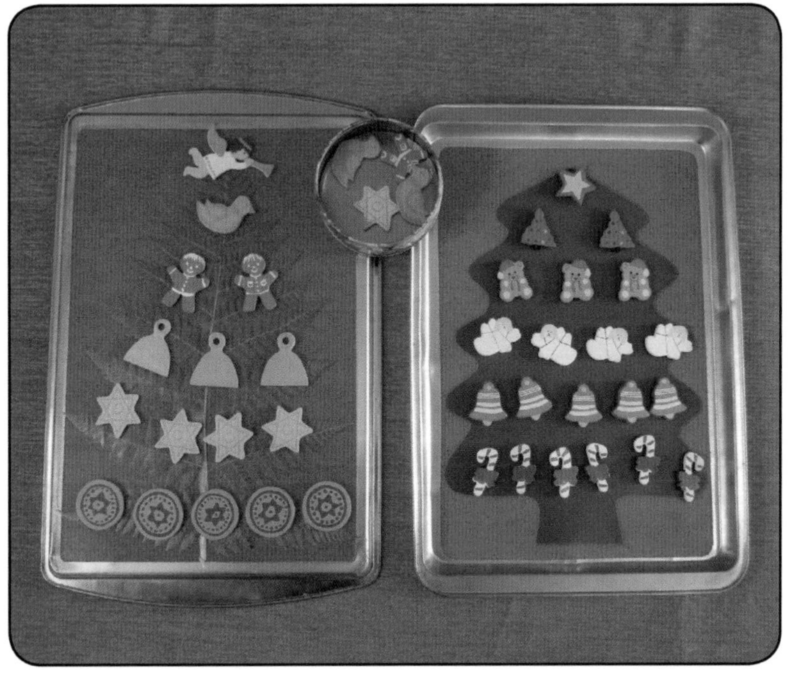

1. Christmas tree: I pressed a fern and put it on red MacTac and covered it with clear MacTac. You can also cut out a tree from green MacTac. Ornaments can be stickers with magnets or anything you can find. It can be a math exercise of counting and sorting.

2. Snowman: Blue and white MacTac for cookie sheet. Be creative with pieces.

3. Old MacDonald: I used popsicle sticks to build a barn first and I sing: "And on his farm he had a barn, with a 2 x 4 here and a 2 x 4 there"...you finish the barn and then start on the animals. In this exercise I added twin geese to make it special for twins that you might have in your class or circle of friends. As you line up the animals it is easier for everybody to sing all the repetitive parts. The fish in the pond adds a bit of fun because there is no sound, you just open your mouth. You can include a bee and go buzz...buzz, etc.

Children like to sing the song to themselves or with a friend as they play with the cookie sheet.

Old MacDonald Song

Young MacDaniel feeding the horse

4. Farm Animals: Counting, sorting, matching. Cover cookie sheet with green MacTac and strips of white MacTac. Add animals.

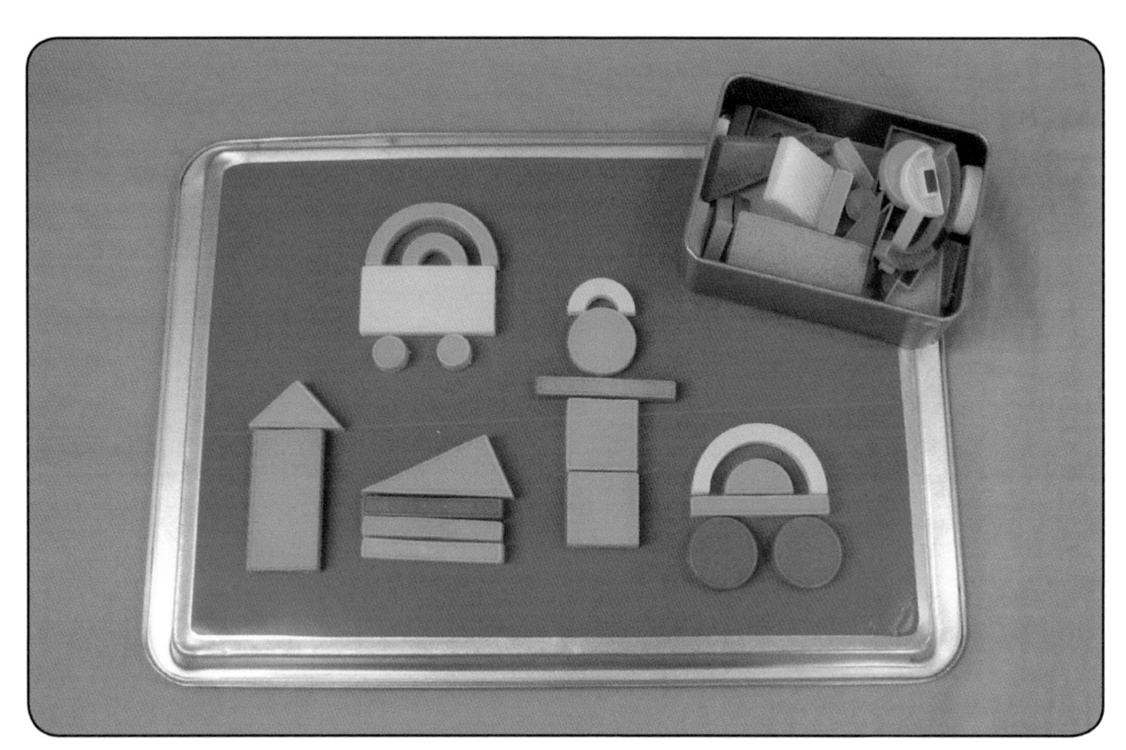

5. Shapes, Buttons & Sticks: Use them to create all kinds of exercises. (See also Page 34). Instead of putting a finished house on the cookie sheet, children have to build it themselves. Holding and naming the shapes and colors helps them to understand the concepts.

Story Baskets for Math & Science

Story baskets are an amazing way to teach math and science to the pre-school child. If children could design a learning style, they would find this method. They love to touch things, lay things out, watch the real thing unfold by their own involvement and imagination. They can count things by touching – it is not just an abstract number. They can hold different shapes, etc.

If you include natural materials such as moss, plants, flowers, rocks, shells, etc, children can touch and explore at their own understanding and speed and it encourages them to explore the outdoors more carefully and find things of interest. In short, it opens their eyes to the world around them.

I believe that self initiated learning for the pre-schooler is the key to a good foundation in later years.

The more we can put good learning habits into the pre-school years, the more we can help them get a good start.

A definition of the Principles of Parenthood by a Persian poet:

- You may give them your love but not your thoughts for they have their own.
- You may house their bodies but not their souls, for their souls dwell in the house of tomorrow, which you cannot visit, not even in your dreams.
- You may strive to be like them, but seek not to make them like yourself for life goes not backwards nor tarries with yesterday.

Let children create some of the props for the stories. Here Oliver paints his own paper boat for an Ocean story..

Story Baskets for Math and Science

Here are the different aspects of learning in each of the individual stories.

Story	Aspect of Learning	Theme	Emotional and Moral Value
1. Life of a Frog A and B	Science Social Safety Math	Pond life, Seasons Day and Night Family life Hibernation Organizing skills	New life Looking out for each other Sequential thinking
2. The Village	Math Job Opportunity Social	Counting, grouping Colors, Matching Delivery service Architecture Family life	Belonging Interaction Feeling accepted
3. The Flower that Smiled	Math Science Social	Counting, sizing Buds growing Butterflies Metamorphisis	Surprise someone Do something nice for each other
4. Grow a Pumkin	Science Weather Social	Seasons Seeds, growing Rainy, sunny	Curiosity Patience Sharing
5 . The Bragging Cows	Science Math	Farm life Cows = Milk Milk products	Bragging Truth comes out Feeling deflated
6. The Big and Little Meadow	Science Math	Farm life Large and Small	Big ego Seeing reality Admitting defeat
7. 5 Little Ducks & 5 Little Bunnies Poems	Math Social Science Poetry	Counting forwards and backwards Ducks - Pond life Bunnies - Forest	Listen to Parents Closeness Belonging Feeling protected
8. 10 Brave Firemen Poem	Safety Math Poetry Dramatizing	Community help Fireman Fire Drill Counting	Feeling safe Being prepared
9. I need an Egg Puzzle	Math Science Social Nutrition	Laying of eggs Baking Farm animals	Being curious Patient Surprised

1A. Life of a Frog

"New Life"

Fabric or Paper Version

Once upon a time, on a sunny day, there was a pond. The water was beautiful and sparkly. This was the pond where two frogs lived. Here they come. They loved to play in the water or just sit on a lily pad in the sun. When spring came, the new leaves on the trees reflected the fresh green in the pond, and the frogs felt a great urge to lay eggs. The frogs were very proud when they saw what beautiful eggs they had laid and watched them everyday.	Lay out yellow cloth. Lay out blue cloth. Bring out two large frogs. Lay out light green cloth. Put two frogs on top. Put eggs on green cloth.
Soon the Spring turned into Summer. The fresh green had darkened to a beautiful green that reflected in the pond. The frogs that were guarding the eggs noticed that they had turned into tadpoles. Now it became very interesting that the little tadpoles were swimming around and exploring the pond. Sometimes they were diving deep down so you couldn't seem them.	Lay out dark green cloth. Leave eggs underneath. Put parents and tadpoles on top.
Soon Summer came to an end and the first leaves started to fall. They floated on the water and looked very pretty. The whole forest around the pond turned color and reflected in the water. It was beautiful. Would you like to see what happened to the tadpoles over the summer? They have all turned into little frogs and each one looked cuter than the other. Now the parents became very busy. They had to teach their children many things.	Put orange cloth on top. Leave tadpoles underneath. Sprinkle on a few leaves. Bring out little frogs.
At night the frogs loved to have a concert going. It was great. The sound was amazing, especially when the moon was out.	Lay out black cloth. Put all frogs on top. Children go "ribbit"..."ribbit."
Finally, the fall gave way to Winter. It got cold and one day there was ice on the pond. The frogs all wiggled themselves into the soft mud at the bottom of the pond to sleep their Winter sleep (hibernate). They dreamed of the Spring that would soon come so that they could again live happily ever after.	Finally, cover everything with white cloth.

Materials Needed for Life of a Frog - 1A

Fabric or paper squares: approximate sizes.

Yellow:	34cm	13"
Blue:	31cm	12"
Light Green:	29cm	11"
Dark Green:	26cm	10"
Orange:	23cm	9"
Black:	20cm	8"
White:	18cm	7"

Eggs, frogs and tadpoles: Copy the sheet with the frogs, once on green paper and once on white paper. Cover with Mactac or laminate, and cut out. Keep organized in a box.

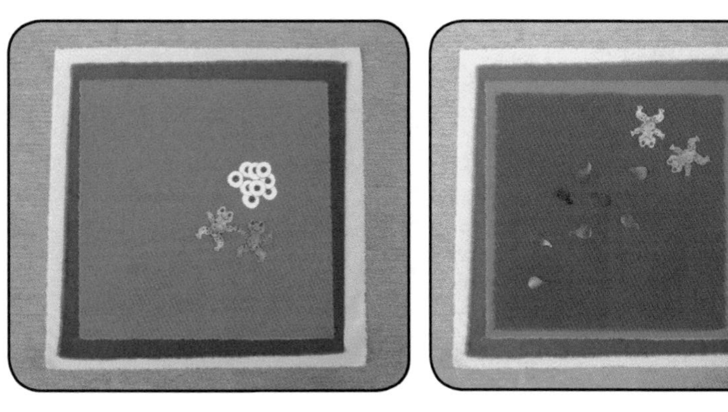

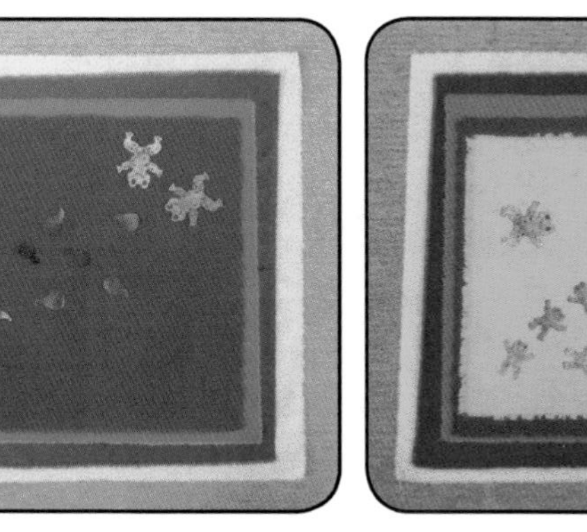

1B. Life of a Frog

“New Life”

Block Version

Once upon a time, on a sunny day, there was a pond. The water was beautiful and sparkly. This was the pond where two frogs lived. Here they come. They loved to play in the water or just sit on a lily pad in the sun. When spring came, the new leaves on the trees reflected the fresh green in the pond, and the frogs felt a great urge to lay eggs. The frogs were very proud when they saw what beautiful eggs they had laid and watched them everyday.	Lay out yellow cloth. Bring box with 9 blocks (blue on top). 2 frogs look out between blocks. Turn 2 blocks to green (lillypads). Then turn all blocks to green and put eggs onto blocks. Frogs sit at edge of box.
Soon the Spring turned into Summer. The fresh green had darkened to a beautiful green that reflected in the pond. The frogs that were guarding the eggs noticed that they had turned into tadpoles. Now it became very interesting that the little tadpoles were swimming around and exploring the pond. Sometimes they were diving deep down so you couldn’t seem them.	Turn all blocks to dark green. Exchange the eggs with tadpoles. Take away tadpoles.
Soon Summer came to an end and the first leaves started to fall. They floated on the water and looked very pretty. The whole forest around the pond turned color and reflected in the water. It was beautiful. Would you like to see what happened to the tadpoles over the summer? They have all turned into little frogs and each one looked cuter than the other. Now the parents became very busy. They had to teach their children many things.	Turn 2-3 blocks to orange. Turn all blocks to orange. Bring out little frogs.
At night the frogs loved to have a concert going. It was great. The sound was amazing, especially when the moon was out.	Turn all blocks to black. Put frogs on top or between blocks. Children go ribbit.
Finally, the fall gave way to Winter. It got cold and one day there was ice on the pond. The frogs all wiggled themselves into the soft mud at the bottom of the pond to sleep their Winter sleep (hibernate). They dreamed of the Spring that would soon come so that they could again live happily ever after.	Turn all blocks to white. Put frogs under blocks.

Materials Needed for Life of a Frog - 1B

Fabric: Yellow 12" by 12" (30cm x 30cm)

Box: To hold nine blocks.

Blocks: You can usually buy nice wooden blocks in craft stores. Paint the six colors blue, light green, dark green, orange, black and white. I painted the blocks so that I can change the whole row together. Then it doesn't take so long when I present.

Eggs, Frogs and Tadpoles: Same as in Life of a Frog - 1A.

Reflections on Life of a Frog

This is a hands-on science story where children learn small motor skills, math skills, and colors, about the four seasons and how to be organized. Children have to think all along and use their memory skills. It encourages enthusiasm and creative thinking. You can make only the one version or both for children to play with. It is the same story but different materials and execution. Because of the staggered nature of the fabrics, children can usually figure out which color comes next. Then they have to match the eggs, frogs and tadpoles to the colors. Children can experience a lot of ability in this exercise. It is not just listening, turning a page, or pressing a button. If we make it special when we present, so that the children can remember for example how the little green frogs looked beautiful on the orange fall color. I include the children by letting them raise their arms as tree branches. In the spring, they would sprout new leaves, and in the fall the leaves would start to fall down, sometimes they would sway in the wind. The little box is like a math puzzle that becomes part of a story. I had a little boy once who didn't think my pond was very good. He obviously found blocks in a box very awkward. Then when the story unfolded, he became very animated with the idea and became quite involved. Later he played with it very often. This is a form of concept learning, a three dimensional play, and great for brain development.

2. The Village

"Belonging to a Family or Community"

Once upon a time there was a very small village with only five houses. One of the houses was red, one was green, one was white, one was blue and one was yellow. They all had a special place for the milk bottles. Every morning, when everyone was still sleeping, the milk was delivered. When the families woke up, the milk was already there. There was one bottle for the red house, two for the green, three for the white etc. Every family took the milk to the kitchen for breakfast. After breakfast, the children got ready for school. It was time for the school bus to arrive. The school bus stopped in the middle of the village and all the children climbed aboard. There was one child from the red house, two from the green etc. Off they went to school together. As soon as the bus had left, the parents planned a surprise for the children. It was a beautiful Spring day and it was the perfect weather to plant some flowers. They wanted to transform the whole village. Along came the flower truck to deliver the flowers. One for the red house, two for the green house etc. Everything looked beautiful. When the children came home, they were very happy, played very carefully amongst the flowers and lived happily ever after.

- Lay out fabric.
- Set out houses and bottle stands to match colors.

- Drive up with milk truck. Put milk bottles into stands according to color. Move milk stand behind house and put children in front of houses.
- Everyone gets on bus and bus leaves.

- Flower truck comes and delivers pots according to color.

- School bus comes, children go to play.

Materials Needed for The Village

Houses: The five colors.

5 Bottle stands: Red - one bottle
Green - two bottles
White - three bottles
Blue - four bottles
Yellow - five bottles

15 Milk Bottles: Red - one
Green - two
White - three
Blue - four
Yellow - five

** color on bottom, otherwise white **

1 Milk truck: Room for 15 bottles

1 School bus: Room for 16 children and one driver.

15 Children: One red
Two green
Three white
Four blue
Five yellow

1 Florist Truck: 15 flowerpots
One red
Two green
Three white
Four blue
Five yellow

Materials Needed for Houses & How to Make them:

- ½" plywood (12 mm)
- Transfer pattern with carbon paper.
- Cut out houses, keep doors., Discard the two small shaded pieces.
- To cut out No. 3, start at tip of roof. Use the little slot above window to fasten a bell. Tie the bell to a string, tie a knot and push knot into slot. Put glue on both sides and sand over it.
- The No. 5 house has two pieces, glue them together. Glue head into window and paint face on.
- Sand all pieces well before you paint.
- Decorate only the front of the houses so that the child knows how to put the puzzle together again.
- You can make a box for everything, lay everything into a rectangle and measure how big the box has to be.

- You can buy finished milk bottles in craft stores.
- Buy molding strips in a lumber store. Profile: 1” by ½” (2 cm x 12 mm). You only have to cut the length for bottle stands.
- Drill holes according to milk bottles. Allow some extra space for paint. You can drill hole right through and glue strips of cardboard underneath.
- I found a little second hand wooden truck that was just perfect to deliver the milk bottles.

Materials Needed for the School Bus:

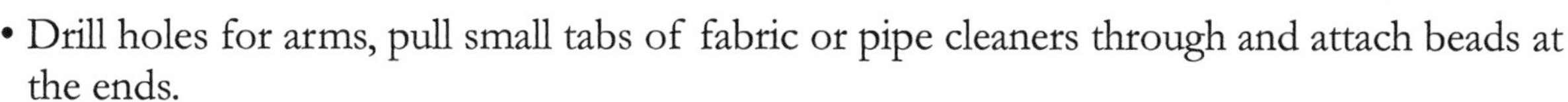

- Cut out bus from ½” (12 mm) plywood.
- Drill holes a little bigger than the people. Sand well.
- Glue cardboard or 1/8” plywood underneath.
- Attach two pieces of dowel on bottom of bus, long enough to attach beads and wheels. The wheels run loose, but the beads are glued on. Some holes have to be redrilled to fit.
- Use round slotted hardwood clothespins for people. Cut at desired length, sand and paint.
- Drill holes for arms, pull small tabs of fabric or pipe cleaners through and attach beads at the ends.

Materials Needed for Florist Truck:

- I found a suitable second hand truck.
- You can buy little wooden flower pots or cut up tiny ice cube trays on the saw. You could also make them with modeling clay.
- Fill these with playdough and use small flowers.
- Be inventive - you might just have something around the house that works. You could use a wagon or a sled.
- I made a rectangular piece of ½” plywood with 15 holes to exactly fit into the back of the truck. Not only do the flowerpots fit, but the children do too!

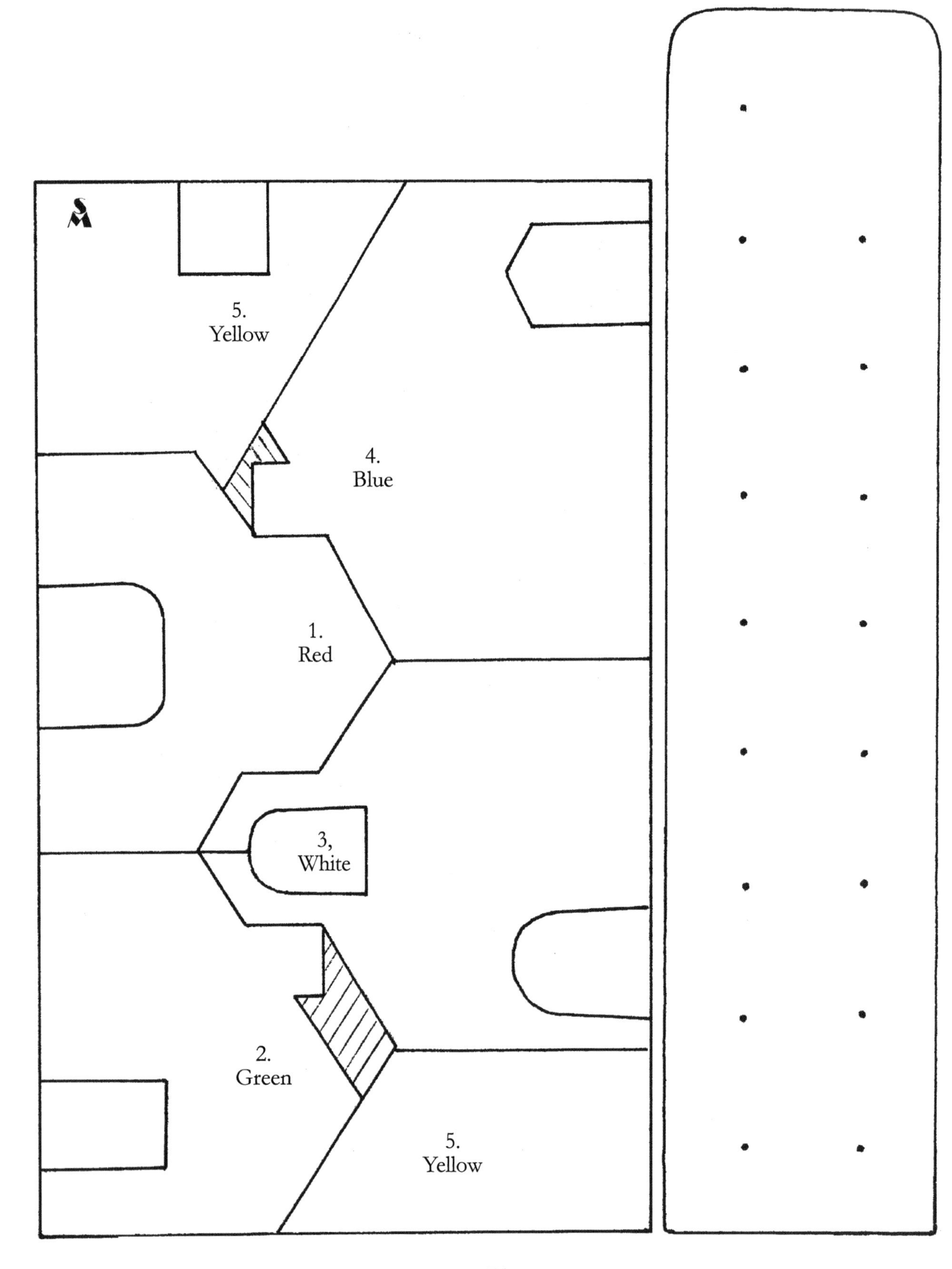
5.
Yellow
4.
Blue
1.
Red
3,
White
2.
Green
5.
Yellow

Reflections on The Village

The Village story can be played in many different versions. When you start, you might want to begin with only the houses and the milk truck. You can concentrate on the colors and the counting. The children can match the color on the bottom of the milk bottle to the color of the house and the bottle stand. They can even pick up concepts for numbers. They can build a pyramid with the bottle stands, and group everything according to number or color. The houses also fit together like a puzzle, so that can be a challenge.

The next day, you can bring out the school bus. Here children can learn more colors and matching and even something about relationships. Some children were very good friends and wanted to sit together on the bus, while another child was sad because nobody wanted to sit beside them. Here you can talk about how important it is to take care of a new person in the class and be nice to someone who doesn't have a friend. Imagine how that child might feel. A child will identify with one of the children on the bus; it might be a favourite color or the expression on their face.

You can also play house, go visit others, invite friends to come over, get the milk bottles, play hide and seek or ring around the rosy (watch out when they all fall down!)

The following day, you can introduce the flower truck. Here it is more confusing because the flowerpots have many different colors of flowers. A child must look at the pot to get the right color to the right house. Soon they will play "park", or bring flowers for birthday's etc. They will most definitely give the flowers or milk bottles a ride in the school bus or take the children for a spin in the flower truck.

It can all be great fun as they learn concepts for math, social skills, verbal skills, small motor skills, colors, creative thinking and roll playing etc.

3. The Flower that Smiled

"Do Something Nice for Each Other"

Once upon a time, there was a little garden, and in this garden, there grew a plant, and in the middle of that plant was a beautiful big bud. Close by lived a beautiful big butterfly who had been watching the bud and was thinking that such a big bud would probably have nice nectar once the flower opened. So, the butterfly kept watching the bud get bigger and bigger. One day the butterfly flew by and said, **"beautiful, beautiful flower, if I do a little dance for you, will you open your petals?"** And the butterfly did a little dance. Slowly the first petals opened, but under them was another tight bud. So the butterfly said again, **"beautiful, beautiful flower, if I do a little dance for you, will you open your petals?"** This time the butterfly circled the flower in beautiful little circles - he sure was a good dancer. Slowly the flower opened more petals, but again there was a tight little bud inside. The butterfly was getting curious and was wondering where the honey was, so he asked again, **"beautiful, beautiful flower, if I do a little dance for you, will you open your petals?"** The butterfly did his very best little dance. As he was watching, slowly the flower opened her last petals and smiled at the butterfly. She had never seen such a beautiful creature and was very glad to let him have some honey. Everyday, the butterfly visited the beautiful flower and they lived happily ever after.

• Lay out fabric, put green felt leaves in the middle, put bud on top. Butterfly flies over flower.

• Butterfly dances.
• Practice the bold parts with the children and let them participate every time.
Repeat this part seven times, be inventive with your dances. You can open two rows of petals at a time too.

• Butterfly gets some honey.

Note: If you twist the string on the butterfly really tightly before you start the story, the first dance of the butterfly will be twirling without you touching the butterfly or moving the stick. It can go until everyone feels dizzy. Everyone will wonder how it moves like that.

Materials Needed for the Flower that Smiled

Garden: Any color or combination of colored fabrics (see photo).
Size 24" by 24" (60 cm x 60 cm)

Green Leaves: Cut out of felt paper or fabric.

Flower: Seven pieces of different colored computer or multipurpose paper. One happy face for the inside of the flower. Copy patterns onto different colors of computer paper.

Butterfly: Paper butterfly on a string and attached to a stick.

How to make the flower: Cut flowers and all the little triangles with a pointy scissors or with an exacto-knife on a piece of cardboard. Fold all the flower petals down so that you have a square. Start with the largest flower, then always glue the next square inside the little triangle cut outs. Study the photo, you will be able to figure it out! You can of course take all the same color, or you can alternate with only two colors.

Reflections on The Flower that Smiled

This very simple story again has something magical. You have the children's attention because they are amazed at the many different layers of petals. It helps the children to pay attention to the many different processes of nature. We should be amazed at something as simple as a bud. How a beautiful flower opens up after being all nicely folded inside a protective cover and getting ready all winter is really quite a miracle. If we as adults do not see it ourselves, then how can children get excited about it? Bud watching in the spring can be quite a fun activity as well as a good learning experience. It is especially fun in your own garden where you can watch the bud develop. Some of the bud's protective covers are sticky to protect the petals from freezing in the winter. In the early spring, the bees collect this sticky substance and use it to seal every little hole in their hives to keep the drafts out and to insulate it.

Horse chestnut trees have big buds that open easily if you cut a branch and bring it inside. You can watch from day to day what happens. They even have a lot of fuzz inside for padding - isn't that amazing? Irises and lilies have big buds that will open when you take them inside. There is surely something growing in your neighborhood that will be suitable.

You can make these paper flowers with your children. For some, you may have to precut the little triangles - be careful with the exacto-knife. On Remembrance Day, I often make poppies this way in only three sizes with red, green and black. Usually we would write PEACE in the center. With my children, we often made these flowers as cards for grandparents and friends with little love messages or wishes inside. We always have a good response to these cards. With one thumbtack you can put one on the wall and brighten up a bathroom or corner.

Flowers and butterflies really outdo each other in beauty and color. Here, I usually use my picture file again to inspire the children for different art projects. We paint butterflies by folding the paper in half and then painting half a butterfly on one side. When the paint is still wet, you fold it in half and like magic, the imprint is on the other side and it is nicely symmetrical. To do this, you can hold the child's hand (who is holding the pencil) and then draw half a butterfly on one side of the paper. Some children love this and end up doing many pictures. We can encourage them to work on color combinations and improve on their work rather than the number of pictures. Cheap paper like newsprint is very suitable. I use some of the excess paintings that children made as wrapping paper for presents.

I always have an art project out for at least three days in a row, because some children will want to do it everyday and improve on certain processes. Others just watch the first and second day to get familiarized with the process. With time, they finally overcome fear and gain the confidence to try something new and they make their own decision to try. You can be there to help in very little ways to make it a success. Often we think we need something new everyday, but we don't.

You can also let the children bring a scarf and have a butterfly dance by holding the scarf behind you at the two corners. You could use Vivaldi's Four Seasons, Spring Movement to dance to. You can have two groups so that the children can watch each other and have more space to dance. Every time the music stops, the next group will get a chance to dance. This is a good exercise to learn group work and to learn to listen and follow instructions. With exercises like this, we help to prepare them for school, even though it may not sound like school, it is very important.

Any dancing to music loosens up the day and gives appreciation to some classical music. Marc, Nisha and a friend feel joy in this exercise.

With this story, we can learn how great it is to do something for each other. A friendship or good relationship always works best if there is an exchange of kindness and not just a one sided effort. It is usually the little things that make a difference. It is the thought that counts. For his 21st birthday, my son got a cupcake with a candle in it from a school friend. He was blown away - what more does it need? We do not need to get repaid for every kindness, but by doing something nice for someone, we might get him or her to do something nice for someone else. I like this kind of snowballing!

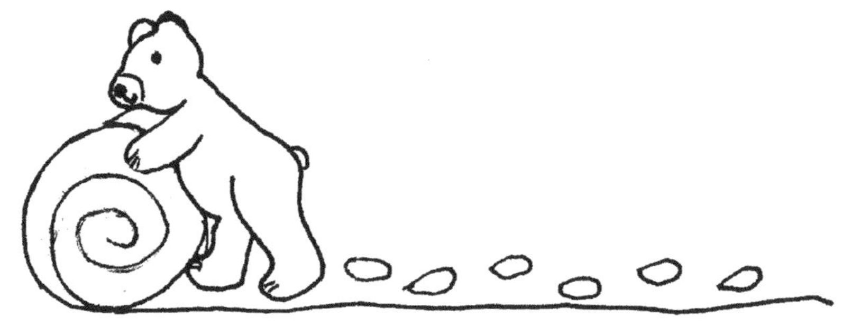

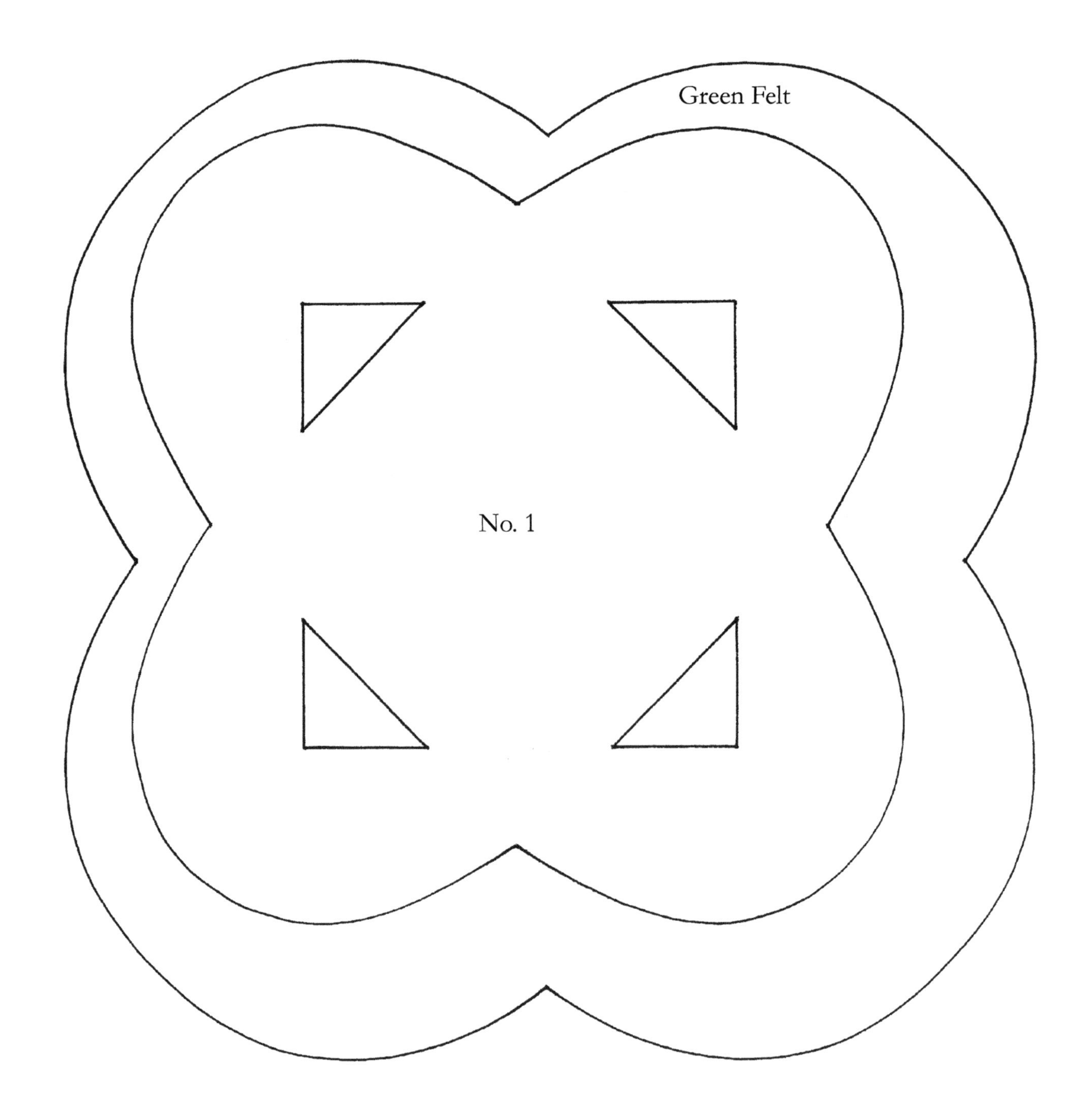
Green Felt
No. 1

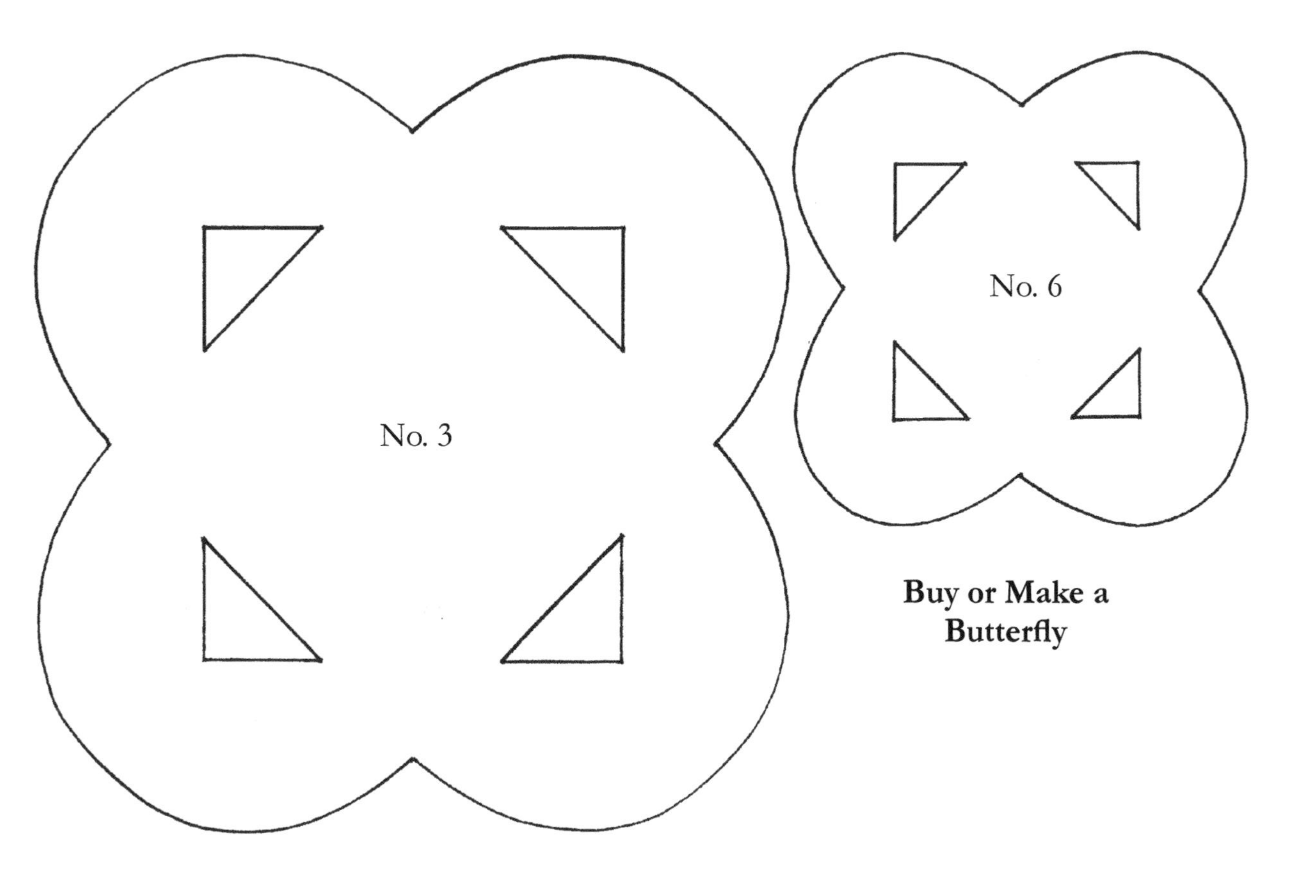
No. 3
No. 6
Buy or Make a
Butterfly

No. 5
No. 4
No. 7
No. 2

4. Grow a Pumpkin

"Patience"

Once upon a time, there was a field. In the middle of that field, there was a little garden. Close by lived a bird who had found a very interesting seed. The bird was wondering what would grow from such a beautiful seed. He went in search for a suitable place to plant this seed, and when he saw the beautiful garden, he planted it right in the middle. Every day the bird came by to see if something was growing already. After a few days, there was a little green shoot peeking through the earth and the bird was starting to get very excited. Now the little plant needed lots of sun. Today the sun was shining beautifully and was bathing everything in a beautiful golden light and warming the soil for the little plant to grow. In between it would rain and the little plant was very happy to drink some water. The whole little garden would be dripping wet and the clouds would make the garden look dark. The rain made the plant grow even more, and when the bird came by the next time he was surprised to see a long vine growing with a beautiful yellow flower at the end. He started to get really curious wondering what would grow on such a vine. Then he got busy with other summer activities. He went on a few fly-ins with his friends and visited some relatives. Then one day he remembered the little garden and flew right over to check on his plant. This time the bird was very surprised to see that a pumpkin had grown and he was even more surprised when he peeked inside and saw so many beautiful seeds. He decided to keep a few for next spring so that he could plant them and have many pumpkins. He shared many seeds with his friends. And lived happily ever after.

- Lay out fabric and little garden. Bring out bird with pumpkin seed in beak.
- Put seed in middle.

- Put out green shoot, bird comes by.
- Cover garden with see-through shiny yellow fabric.
- Cover garden with blue tulle.

- Put out vine with flower and bird comes by.

- Put out pumpkin with seeds inside, bird comes by. Bird looks into pumpkin.

- Bird reflects.

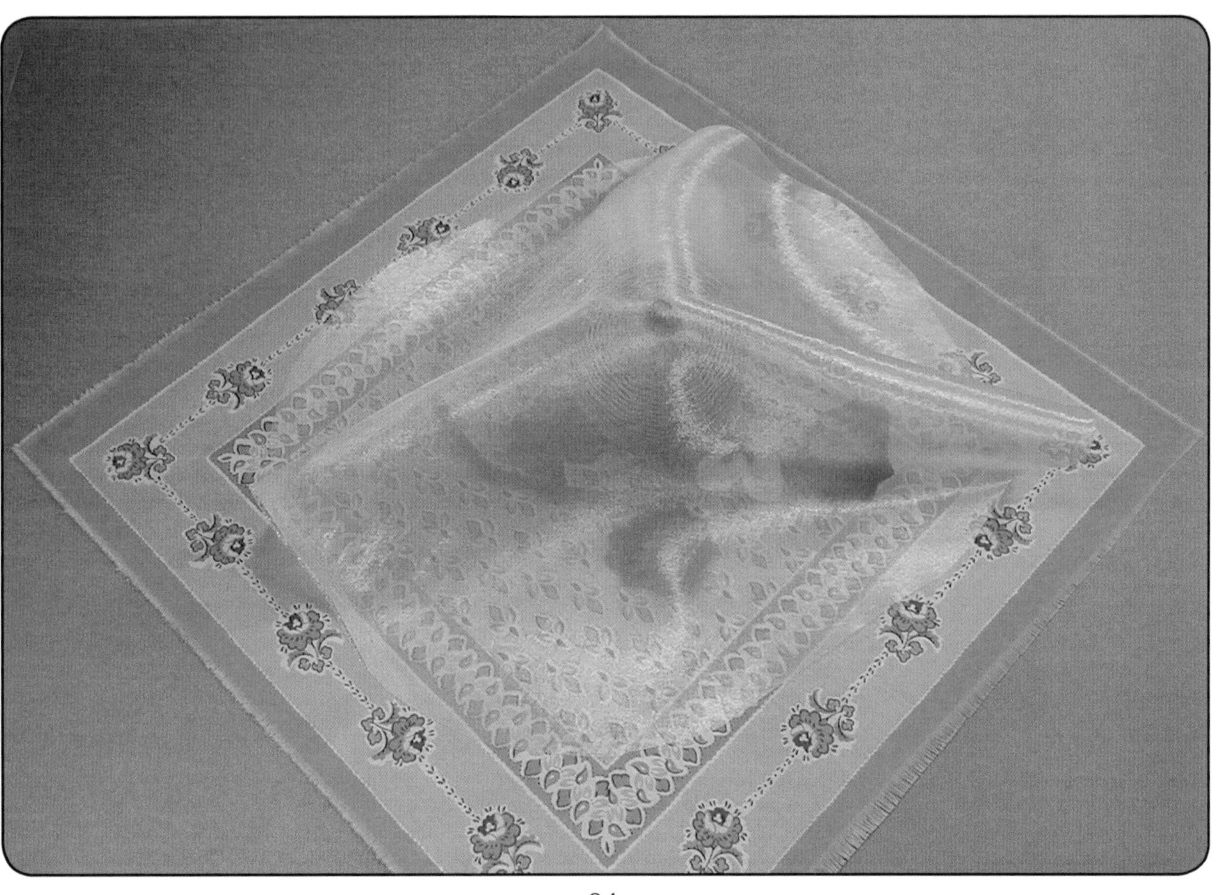

Materials needed for Grow a Pumpkin

Fabric: 1 piece 24” by 24” (60 cm x 60 cm) earth tones. 1 smaller piece to fit as a diamond on top with flowers to look like a little garden. You can paint something yourself or take wrapping paper. 2 pieces of see-through yellow and blue to be the sun and the rain.

Pumpkin: Take a real one, a wicker basket or a plastic one.

Pumpkin Seeds: Preferably real seeds

Bird: Cut out of ½” plywood or take any bird.

Vine: 2 green pipe cleaners and yellow and green felt.

Reflections on Grow a Pumpkin

“Patience”

This story can be used for other vegetables or flowers. For example: with corn or with a sunflower. Once, I had a huge sunflower with lots of seeds which was very interesting for the children. It is valuable for the children to have the growing season explained this way. I always involve them by asking them to be the sun when it shines or to make the movements of rain when it falls. Children are more able to concentrate when they can move a bit in between.

I designed this story to be a Halloween story. You can use a real pumpkin and then use it as a jack-o-lantern. I would already have it cut open on top and the seeds washed and put back into a clean pumpkin. Often, I would roast the seeds with a little salt and we would eat them the next day.

This is also a gentle story for Halloween where nobody will get scared. This story again connects children to real life. Every time they see a pumpkin, they will know how it grew and that it takes a long time. The bird came back many times and had to wait a long time until it had finally grown. We can be excited about waiting, instead of being impatient. It is good to learn about patience because impatience does not make anything go faster and mostly just makes us miserable. These are all opportunities that give the children a feeling of time. They learn by experience how long something will take. They are mostly impatient because they don’t yet understand the time concept.

Just to talk about seeds is another great project. Every plant has a flower, a fruit and a seed. Some seeds fly away in the wind, some stick to animal fur, some are large, others are tiny. Many we can even eat. You can point out the seed shelves in a store or even better you can let the children make their own collection. You can give each child an egg carton and when they find some seeds, they can glue each variety into a different compartment. You will see right away that some children will be very interested and will really work at it while others will forget about the project quickly. With activities such as this, children will be able to develop different interests.

To plant seeds in little pots or in a garden would of course be the ultimate experience. There, children will really experience how long it takes and how exciting it is when something finally peeks through the earth.

Two Cow Fables and One Cow Poem in a Little Box

5. The Bragging Cows

"Be humble"

Once upon a time, there was a meadow, and in it there were three cows eating hay. The first cow said, "I can give so much milk that every little child in the world can eat ice cream". "This is nothing," said the second cow, "I can give so much milk that a whole river of milk can flow to the city."	Lay out green fabric. Use the lid of a box or paper manger with a little hay. Bring out three cows.
The third cow said, "I don't have that much milk. I only have a few bottles everyday."	
Along came a little sheep who was very sad because he had lost his mommy. He had an empty bottle and asked the cows if he could have some milk. The first cow hung her head and had to confess that she had no milk. It was the same with the second cow. The third cow gladly gave some milk to the little sheep. She filled the bottle to the very top. The sheep thanked the cow and ran away happily.	Bring out sheep with milk bottle. Cow fills bottle

Materials needed for the Bragging Cows

Fabric: A nice green cloth.

Cows, sheep and Milk Bottle: Anything you can find, craft stores have unpainted, wooden cows for you to paint, also sheep and milk bottles. Use any farm animals you can find.

Hay: Cut a little bit of lawn with your scissors and put on a tray to dry. It is great to use fresh hay.

Box: I have a little tin can in which everything from the story can fit and the lid can be used to put the hay in for the cows to eat. You can make a very simple manger with a piece of cardboard. Study the photo.

Reflections for The Bragging Cows

We often feel like covering up some shortcomings with a bit of bragging!! Feeling inadequate or inferior can trigger some need for boasting. Experience will teach us that being found out is much worse. Without adding any explanation for this story, children will clearly see how embarrassing a situation like this can be. But best of all is when we realize that we are fine in who we are. We are going through productive and dry times and we can be useful in every phase in out lives. We don't have to "give milk" to be special.

6. The Big and Little Meadow

"Don't think of yourself as more important than others."

Once upon a time, there was a big meadow in which Rosy the Cow was crazing. Next to it was a small meadow where Frieda the Cow was also grazing. Frieda in the little field thought, "I am much bigger. Look how small the other cow looks in her field." And she said to herself, "I have a much bigger appetite. Look how short I keep the grass in my field." And she chuckled to herself when she thought about how much faster she could walk. It took her no time to walk all around her field. Rosy, the other cow, took a long time to walk around her field.

Put out large green fabric and small green fabric.
Bring out cows.

One day, Rosy and Frieda met in the corner of their fields and Frieda was very surprised that she was not bigger than Rosy. When she was invited to Rosie's field, she was surprised to see that even with her help of eating grass, it was hard to keep it short. And as for a walk around Rosie's field, that really taught her a lesson. She was out of breath before she reached the end. Frieda swallowed her pride and they lived happily ever after.

Put Frieda in Rosie's field.

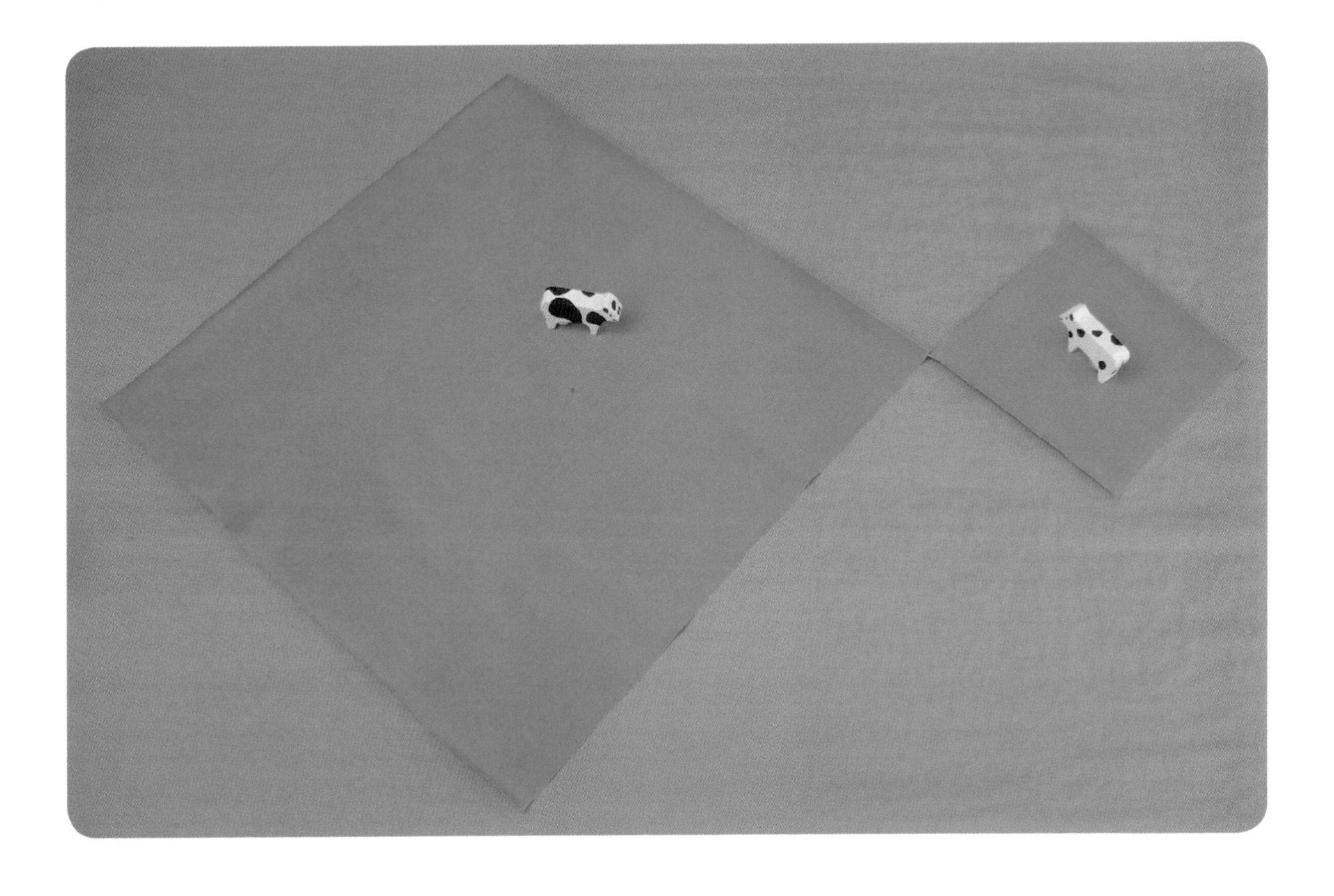

Materials needed for The Big and Little Meadow

Fabric: 1 large and 1 small piece of green fabric according to the size of your cows.

Cows: Anything you can find (it could be horses too). They have to be the same size.

Reflections on The Big and Little Meadow

Somehow, we see everything from our own standpoint and often it doesn't give us a completely realistic and true picture. It is probably partly combined with wishful thinking and a feeling of self-importance. All these feelings are often very close neighbors of insecurity and inadequacy. Especially in our childhood we have to sort out a lot of feelings and any help we can give with stories and their moral is great.

We can be mistaking, misjudging or miscalculating. We can be misinformed or misunderstanding a situation, we don't have to be always right. We can learn to admit our mistakes and realize how liberating it is. We don't have to know and be an expert in everything.

A Poem to go with your Farm and Cow themes

"THREE LITTLE COWS"

Three little cows want something to eat — Moo moo, moo.
Hay or corn or grass would be neat — Moo moo, moo.
What will the farmer feed them today — Moo moo, moo.
I think the mangers are filled with hay — Moo moo, moo.
The farmer is asking for milk with a please — Moo moo, moo.
Because he wants to make some cheese — Moo moo, moo.
He says the children love butter and cream — Moo moo, moo.
But most important some yummy ice cream — Moo moo, moo.

As you say the poem, clap your hands and let the children say "moo moo, moo."
Keep a good rhythm.

7. Ducks & Bunnies

"Listen to Mom & Dad"

In this section I would like to point out that you can not only use stories, but poems and songs as well. It is nice to introduce a poem with props and children love to set up little areas with animals, trees & flowers. As you can see in the photographs you can find suitable wicker animals that make a nice toy, if you can add a soft fluffy mat and the babies.

You can have a variety of ponds, lakes and meadows in different sizes and materials. You can cut them out of 1/4" plywood, posterboard, plastic place mats or fabric.

All the animals and trees are cut out from ½" (12 mm) plywood. If you like you can also use the type of tree I introduced in the Magnetic Stage section (Book 1, page 60).

These sets make nice gifts or they can be taken along on a trip or when you visit friends, so the child has something to play with.

Animals and trees can always be used in the block corner or with the train set etc. Design your own cows, geese, pigs etc.

Flowers:

Have a basket with flowers for meadows and borders. Children enjoy setting them in little rows or groups. They are very easy to make. You probably already have a basket somewhere!

Use ½" (12 mm) dowel and cut it into different length pieces. You can easily use a handsaw. Paint them green or leave them natural wood. Cut out felt flowers and glue them on, then put a small upholstery nail in the middle.

Five Little Ducks

In the evening on the pond,
where the ducklings swim round and round.
The mother calls them to the nest,
so the little ducks can finally rest.
And under her wings they happily creep,
as she quacks a song 'til they fall asleep.

Five Little Bunnies

Five little bunnies come out to play,
the first little bunny likes to sit in the hay,
the second little bunny says let's play hide & seek,
the third little bunny wants a carrot to eat,
the fourth little bunny hops around a wild rose,
the fifth little bunny wiggles his nose,
but soon mother bunny calls them to rest,
in a beautiful, warm and cosy nest.

8. Fireman Poem to act out with props for Dramatizing

"Be Ready"

We all have to do fire drills in our classrooms and it is even good to do them at home so everybody knows what to do in an emergency. Often children get scared, but I found that if children can be a fireman they respond more favorably. So I made a fun poem and a contraption that goes with it. It is a fire engine or a pole where the firemen (beads) can slide down.

Ten Brave Firemen

Once upon a time there was a fire hall, And there was a fire chief. In the fire hall were ten beds. And there were, 123456789	Lay out red fabric. Put on fire hat. Put out strip of beds. Bring out 10 beads on pole. And start the poem by counting and moving the beads. Put firemen on pillows.
Ten brave firemen, Sleeping in a row. Ding dong goes the bell, Down the pole they go.	Ring bell. All slide down pole.
Jumping on the engine, Off they go. Putting out the fire, To the very last glow.	Put ladder on fire truck. Drive off. Make motion of putting out a fire.
Tired and content, To the fire hall they go. Back to bed again, All in a row.	Drive back to fire hall. Put firemen back to bed. Keep a good rhythm

By looking at the picture you can put something together very simply yourself. I used large beads, drilled the holes a little bigger to fit a (thickness of a pencil) piece of dowel large enough so there is space to slide down. The beads are confined so they don't roll all over the place. It acts like a counting stick also,

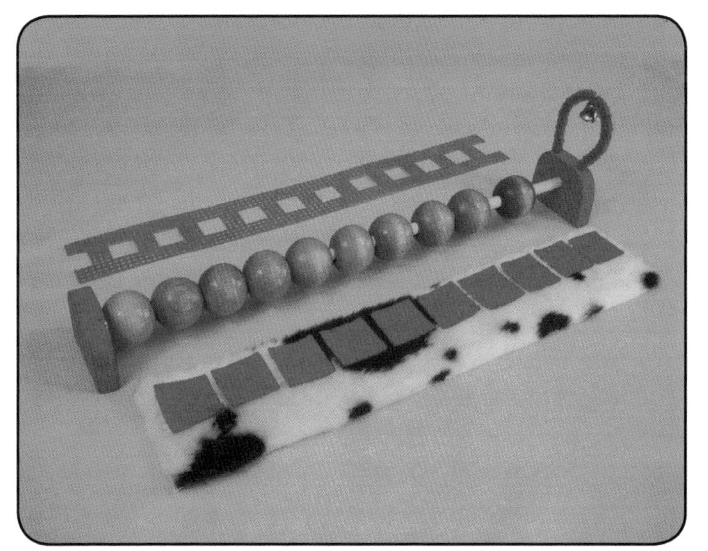

When you dramatize, have a row of little chairs, as many as you have children, it can be 3 little firemen or 12, etc. Little mats for beds, a bell and pretend movements and away you go.

9. I Need an Egg Puzzle! - "Be Fascinated"

Story Box - 9 layer puzzle box with 1/4" (6 mm) plywood.

Precut: 9 squares 1/4" (6 mm) plywood. 3" by 3" (7.6 x 7.6 cm)

Box: 1 bottom sheet 1/8" (3 mm) plywood - 4 pieces of molding as in Box 1, Page 102.

Trace patterns onto squares, cut pieces on the saw, sand and paint. Each square has only three pieces except the egg which has four. The box is used as a cubby hole with a little bit of hay for the chicken to sit in to lay it's egg. I have taken the idea of the story from a nursery rhyme that I grew up with. You will recognize the same idea in my puzzle, "Benjy and the Apple Tree." in Book 4, page 71. This is a story that has a lot of repetition that the children like very much.

This is the story: Lay the box on it's side, put some real hay inside and put the chick and the egg into it so that you cannot see the chick. Next, line up the mother, the boy, the dog, the cat, the pig, and the rooster. It starts:

Once upon a time there was a mother who wanted to bake some cookies. She didn't have any eggs and she needed to make some dough. So the mother sent her little boy down to the chicken house to see if the chicken had laid an egg. When the little boy came down to the chicken house, he saw the chicken sitting in the hay, but the chicken did not want to give the egg to the boy. The boy was fascinated and stood there watching the chicken. Now the mother who was waiting for the egg was wondering why her little boy did not come back, so she sent her dog to check out what was happening. The dog was also fascinated to see that the chicken did not want to give away her egg to the boy and stood there watching the chicken. So the mother sent the cat....and then the pig... and finally the rooster because she knew that the rooster would know his chickens. When the rooster didn't come back either, the mother decided to go and look for herself. When she came down to the chicken house, she was amazed to see what was happening. The chicken started to cluck loudly and was quite happy to have an audience. As everyone was watching, the chicken came off the egg, the egg cracked open and a brand new little chick hatched before everyone's eyes. It was a beautiful sight and everyone was happy and admired the fluffy little chick. The mother decided to find a recipe that didn't need any eggs and they all lived happily ever after.

This story would be suitable for a story basket too. You can take a cardboard box as a chicken coop with a bit of hay and use any animals you might have. It could be a cow, sheep, goat etc. The word fascinating is really too difficult for small children. It is nice to use a different word once in a while and this one gets repeated a lot so the children can learn it quickly. This is how children learn new words. They can feel a little important by being able to say such words. When the story is finished, the children can put all the pieces back into the box. When they play on their own, they usually play the story too and will have to recall the sequence of the story

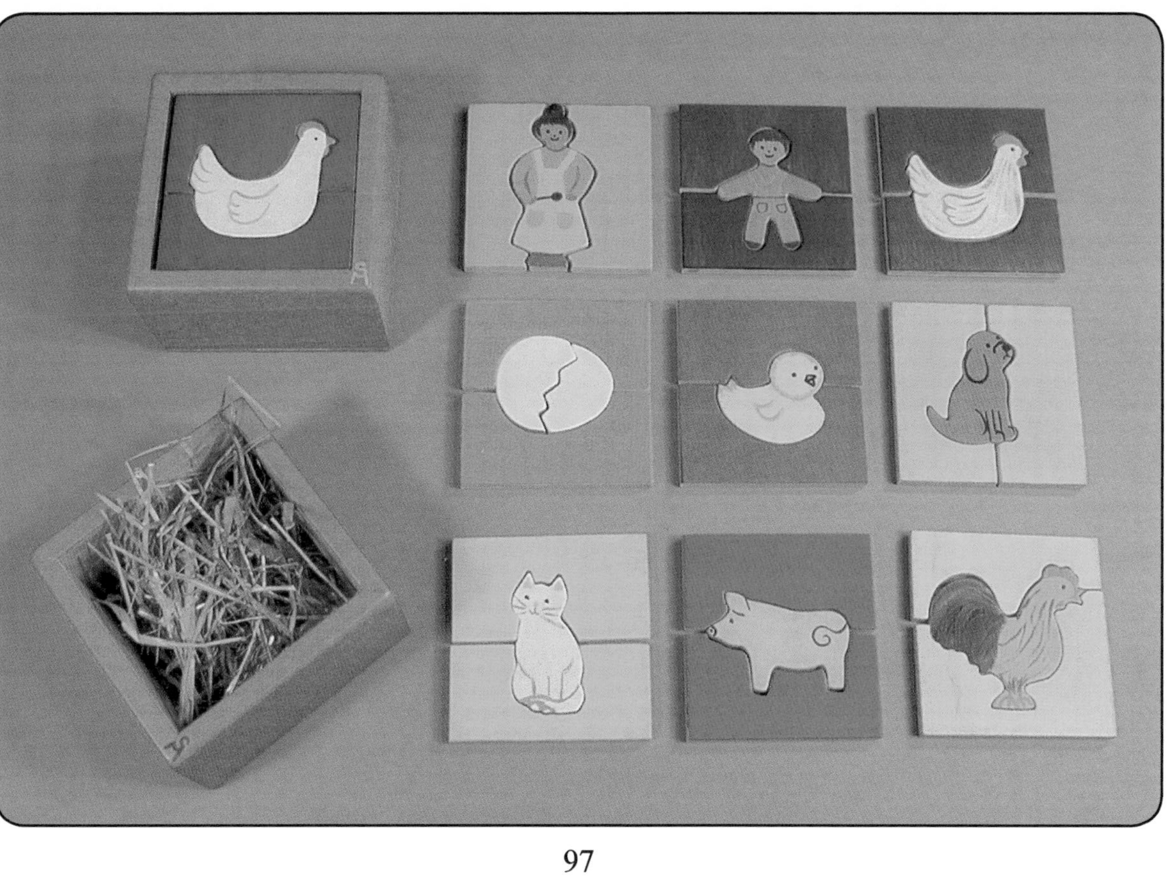

Preview of Book 4 - Puppets & Empathy

In this part we look at puppets and how they can be used in education. Children can easily connect and relate to puppets which allows them to present amazing ways to teach virtues and empathy. We look at different varieties of puppets and show how to make them. Included are 7 more story baskets focussed on empathy and a discussion on play and playrooms.

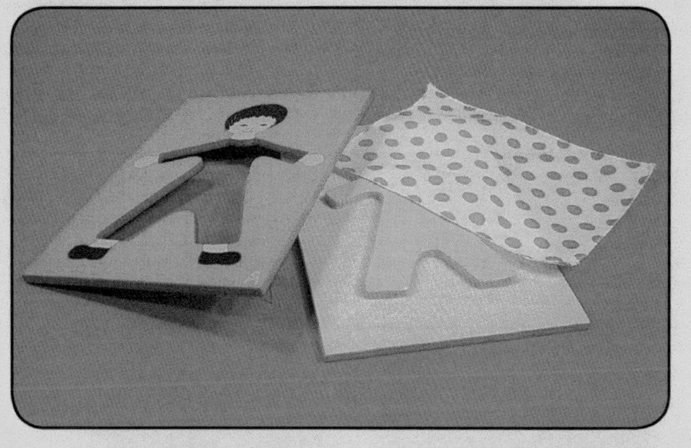

Tools and Materials

I should stress that I realize that not everybody who reads this book series is necessarily the type of person who will go ahead and make all the projects presented. Some readers will take the concepts which are presented and use them to buy good toys for their children, whereas other readers will go ahead and make the puzzles, puppets, etc which are presented.

However, as mentioned earlier on in this book, everybody can make the projects, and I hope that all the readers will try and complete some of the projects. Making a toy for one child or a group of children must be one of the most satisfying experiences I know. When you watch a child's face as they play with your toy and see them learn, it is priceless!

In this section I present some tips and suggestions for those readers who would like to make some of the presented projects. I have developed these over the last many years through trial and error.

I should also mention that besides purchasing the specific materials mentioned in this book from your local hobby shop, you can also use scraps of materials that you find around home. You can visit your local carpenter, furniture or model building shop for scraps. In many cases the use of scraps can lead to inspiration and creativity.

Wood

Plywoods:

I always get good quality plywood for many reasons:

1. It has less slivers.
2. It cuts nicer on the saw.
3. It is a lighter color and the paints look brighter on it.
4. It is stronger.
5. You have less waste.
6. The finished toy looks much nicer.
7. It is easier to work with.

It is worth it to pay a little more for the better quality plywood. Finished plywood that you use for toys has to be good on both sides. I always make sure that the better side is on top of the finished toy. Once you determine what quality plywood you want, you still want to check all the sheets, because they are not all the same. You can save a lot of time later when you sand and finish, if you take care when selecting the sheets. Most of the plywood sheets I use are birch wood. I use the following thicknesses:

1/8" (3 mm)..................... 3 ply
1/4" (6 mm)..................... 5 ply
3/8" (9 mm)..................... 7 ply
1/2" (12 mm)................... 5 ply

You can get ½" plywood that is 9 ply (meaning that there are nine layers of wood glued on top of each other), but that is very hard to cut with the saw. The ½" plywood that I use has three layers of fairly soft wood on the inside. On each side there is good quality hardwood on the outside, so it is 5 ply. It cuts very nicely with the saw. I usually have the plywood sheets cut into strips at the lumber store, so that I don't have to handle large sheets. Most of the lumber stores will sell you ¼ of a sheet. When you only make a few toys, you do not need a whole sheet.

Moldings:

I use different types of moldings for the boxes. They are a good quality wood and there is a good variety. You will have to see what you can find at your lumber store. I mainly use three different types. Choose good strips without cracks or slivers.

Make sure you design your boxes a little larger than your squares. I like to have about 1/8" (3 mm) extra space. If the pieces fit too tightly into the box, it can be frustrating for the children. If there is too much space the puzzle doesn't sit nicely.

Box 1: This molding fits nine layers of plywood. As you can see it is rounded on one corner, it makes nice boxes.

Box 2: This one fits five layers and is also rounded. I cut this one down for my Four Seasons Apple Tree (page 45). The height of this puzzle is three layers of ¼" (6 mm) plywood plus one layer of 1/8" (3 mm). For the Christmas Tree puzzle which is two layers of ¼" (6 mm) plywood plus one layer of 1/8" (3 mm), I cut it down also (Book 4, page 61).

Box 3: This molding is ideal for three layer puzzles and for larger boxes. I use this one for the hammering game box too (pages 32 & 33).

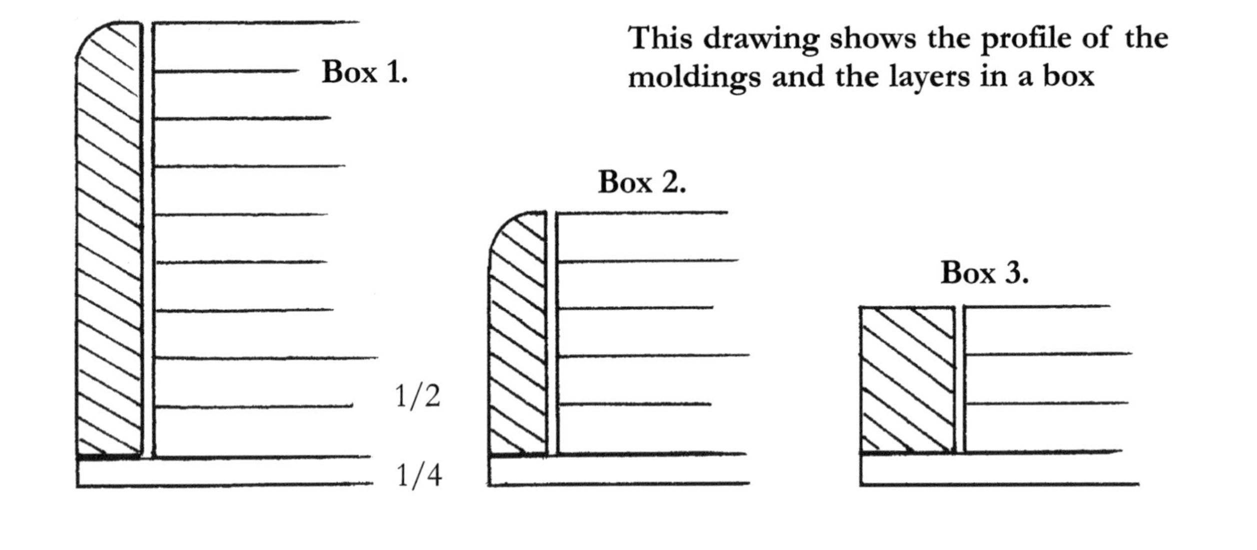

This drawing shows the profile of the moldings and the layers in a box

Dowels:

I used many different thicknesses of dowel for the different projects. You will see them throughout this book series with such projects as:

* finger puppet stand
* broomstick dolls
* peg boards
* flowers
* cone puppets
* hammer
* handle for magnetic stage

You can buy hardwood dowel in many different diameters that are quite expensive. There are also some standard diameter softwood ones that are more reasonable.

Tools and Materials

Table saw:
I have a small table saw in the garage, where I precut all the parts for the boxes and all the squares and rectangles for the puzzles. When I set up the saw, I usually cut a few pieces at the same time. Even if you make two the same, you can save time and make nice gifts.

Fret Saw:
This is a hand saw. It is very reasonably priced and works great for ¼" and 1/8" plywood. For the first few years, I cut everything on this saw. It has a very thin saw blade (like a wire with teeth) so that you can cut tight corners (See picture above).

Scroll Saw:
This is my electric saw and I wouldn't know what to do without it. I use it in my work room, where I have everything in easy reach from my swivel chair. There are many different saws that you can buy. One thing to look for is the attachment of the saw blade. My saw has very strong saw blades with little pins already attached so that I can easily exchange them. It is very frustrating, if you have a saw on which the blades always break. It is also good to have a built in vacuum so that you do not breathe in too much sawdust. I always wear ear protection and a dust protector mask. If you do regular maintenance on the saw and do not use it for hours at a time, it will last a long time. I usually cut for about ½ hour to one hour and then do

something else. Maintaining the saw means oiling it regularly and blowing out all the sawdust, especially in the motor, when it becomes necessary. You do not want it to clog up.

Paints:
I use acrylic paint in 2 fl. oz. (60 ml) bottles. I also use one or two sets of these paints for workshops depending on the class size. You can find them in stationery stores, hobby and craft stores, lumber stores, department stores etc. Shop around for a good price. There are many different brands but so far I have been able to mix them together even if they come from different companies. The consistency is usually just right for these kinds of projects. Sometimes, you have to add a little water. These paints are non-toxic and waterproof so you will never get dizzy. You can wash the brushes with soap and water. I squeeze out a bit of paint on a plate. This way you can easily mix the paints if you have to add a little white or any other color. Brush the paint on evenly (not too thick and not too thin). You will soon get a feel for how it works best. For most of the colors you only have to do one application, though white, yellow or green often need another coat. I brush the paint on with the grain of the wood. I always put the base color on first and then wait (approx. 15 min.) before I paint on the details like faces, flowers, and special lines.

Brushes:
I use soft brushes from ½" (12 mm) down to very fine ones for detail. Brushes are expensive, but if you take good care of them, you can use them for a long time. If you do not wash them right away, they get hard and are ruined. I always wash them with soapy water.

Varnishes:
You do not have to varnish these paints, but it gives it extra protection and a better finish if you do. I use water based Varathane. There are three different types; clear gloss, clear semi-gloss, and clear satin (or flat). Use the one you prefer. I mainly use semigloss and satin, but gloss looks nice also and is really good for when you have to wipe the toys with a damp cloth once in a while. I prefer brushing the varnish on. It is always a nice finishing job. If you spray varnish, make sure you do not get the pieces stuck to the newspaper.

Glue:
For all my projects I use white glue. It works especially well for wood. It bonds fast and dries clear. You can buy the white school glue; it is non-toxic and washable. I wipe excess glue off with a slightly damp cloth. For projects with photos or calendar pictures, I use glue sticks so that it bubbles less.

Putty:
If you have to fill holes in the wood, you can use Durham's water putty. It is a powder that you mix with water to the right consistency. Fill the holes, let it dry one hour, sand over it and then paint it. It sticks well and does not shrink. Sometimes there are holes in the plywood and when you cut out your pieces you come across them. This putty is good because you can mix small amounts of putty with just a few drops of water.

Nails:
Keep a small assortment of nails handy. To finish boxes, I use ½" (12 mm) finishing nails.

Final Thoughts

I think the great motivator for me to do this project is that I carry a bit of a burden for the children. On the one hand we live in an incredible world of opportunity, on the other hand we are socially and economically pushed into a mold, which bears at lot of frustration and pressure. We all suffer, but the children suffer greatly. The expectations and stress many of our children and teenagers live with is far above a normal level anybody should bear.

Since we are part of this world and system, it is my aim to help the children develop some coping skills and acquire concepts for learning as well as learning more about who they are, their talents, likes and dislikes and so become more secure within themselves. Children love and need the feeling of "I understand," "I get this," "I am good at this," "I like this," etc., instead of the insecure feeling of being lost and not knowing where they fit in.

I know a lot of people feel like I do, and we need to support each other to at least make a little bit of a positive difference. Unfortunately, it takes great effort to swim against the mainstream, but it is very worthwhile, if we can even help only one family or child.

We all know that the right amount of discipline gives the children a secure feeling. Education is really a balancing act between discipline and love. Unfortunately, there is no formula, each child responds, learns and develops at different speeds and levels. We have to constantly adjust, adapt, be flexible and creative to try to stay on top of things. This needs a lot of energy and only too often we don't have enough of it, we are exhausted, turn a blind eye and give in.

But this is life, we are never perfect and we don't have to be, as long as we try our best, have a positive attitude, practice patience and enjoy our children.

They are our greatest heritage, they will be our next generation and so determine the future.

About the Author...

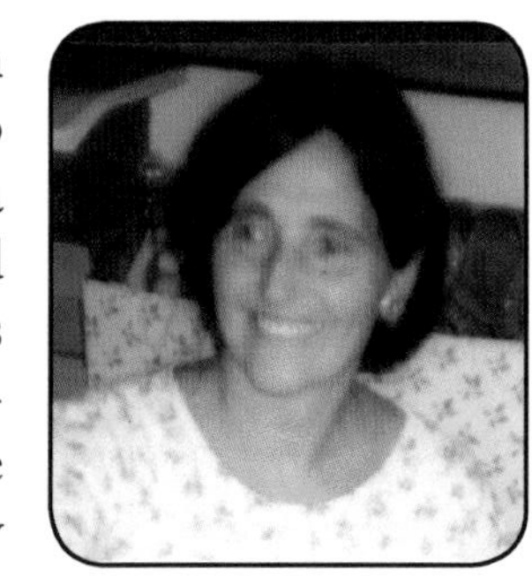

I was born, raised and got my Degree as a Kindergarten teacher in Switzerland. 40 years ago, Eric and I were married and emigrated to Canada. I worked in the ECE field ever since, either as a teacher, a curriculum designer, a conference speaker or as a Mom to Marc and Nisha, now 34 and 32. 25 years ago I started to give workshops throughout the Vancouver area to ECE workers, students and to various parent groups and worked at the local University College of the Fraser Valley part-time as a sessional in the ECE department for a few years.

Children are my passion and toys are my hobby.

After much encouragement I have finally put my ideas, thoughts and experiences into "**Learn to Play - Play to Learn**."

It is my belief that the potential within the child is much greater than we could ever imagine. **It is my hope** to make a difference for the children and to awaken a new love and excitement for all people working with children. It is truly one of the greatest jobs to raise our next generation. **It is my wish** to create an awareness that too much TV watching is not only unwholesome for children but squanders a lot of valuable time at a critical age that will never return. We **can** make a difference that will help build a good foundation for life.

I have a special love for parents who make a commitment and take child rearing seriously. If I can help put more joy, fun and at the same time some concept learning into this process for them I will have succeeded. I can boldly say I loved being a Mom. I was very privileged that I did not have to work full-time outside of the home, and was able to keep in touch with my profession.

With this book series I hope to put more importance on child rearing and encourage young mothers (and fathers) to stay at home, especially in the early years if at all possible. I am very encouraged to see a strong movement back to this and hope that eventually society will give it the importance it deserves.

The combination of my own childhood, raising our own family, the thorough education Switzerland gave me, the great opportunity Canada provided me with, as well as God's unfailing guidance brought this book about.

Made in the USA
Charleston, SC
06 January 2013